, Moriz Benedikt

Brains of Criminals

, Moriz Benedikt

Brains of Criminals

ISBN/EAN: 9783741182624

Manufactured in Europe, USA, Canada, Australia, Japa

Cover: Foto ©Andreas Hilbeck / pixelio.de

Manufactured and distributed by brebook publishing software (www.brebook.com)

, Moriz Benedikt

Brains of Criminals

Harvard University

Library of
The Medical School

The Gift of
THE NEW ENGLAND
DEACONESS HOSPITAL.

ANATOMICAL STUDIES

UPON

BRAINS OF CRIMINALS

A Contribution to Anthropology, Medicine, Jurisprudence, and Psychology

BY

MORIZ BENEDIKT
Professor at Vienna

TRANSLATED FROM THE GERMAN

BY

E. P. FOWLER, M.D.
NEW YORK
Department of Translation New York Medico-Chirurgical Society

NEW YORK
WM. WOOD & COMPANY, PUBLISHERS
27 Great Jones Street
1881

DEDICATION.

TO THE

CENTRAL DIRECTOR OF THE ROYAL CROATIAN PRISON,

EMILE TAUFFER;

AND TO THE

PRISON SURGEONS,

DR. BETTELHEIM (LEOPOLDSTADT, HUNGARY),
DR. BADIK (ILLAVA, HUNGARY),
DR. ROHAČEK (LEPOGLAVA, CROATIA),

THIS WORK IS DEDICATED.

DEDICATION.

The dedication of this work seems due to you, as it was through your disinterested and self-sacrificing assistance that I was enabled to furnish the foundation stones towards a Natural History of Crime.

That Hungary, my native land, and Transleithania in general, should especially have furnished me most material assistance was certainly not an accidental occurrence, but one which might reasonably have been expected.

There it has been an immemorial custom to hold fast to one's convictions and to give them unreserved expression, even though exposed to the danger of a temporary extreme defeat.

It is true that my desire to obtain material for my studies was responded to in a most friendly manner by the Supreme Court of Vienna, and by His Excellency the Minister of Justice, Dr. Glaser; the latter indeed is too eminent a scholar to undervalue the importance which must attach to these studies, even though the results might be negative.

The monstrous counter-agitation infused throughout the educated classes by the Vienna Academic circles, rendered it virtually impossible, however, for me to profit by the kindness of the Chief Department.

This was of course a serious obstacle, for in Cisleithania alone I could have chosen my material from a single race.

First of all, I must here express my warmest thanks to

Professor Betz, of Kiew, who was at the same time a spur, a guide, and a support to me. It was only the greatest confidence in his authority and special knowledge in this branch that encouraged me to continue my studies in spite of the great distrust which they encountered, and to conquer the subjective fear of hopelessly wandering around a source of error; a fear which must necessarily possess every one who stands isolated with his facts.

The lion's share of the labor, though, has fallen to the honored investigator who bestowed every possible attention upon the outline-drawings, and superintended the technical execution of the plates.

I must also return my most profound thanks to His Excellency Baron Anton von Hye. He is a scholar of the good old Austrian school, who, surrounded by the influences of the mighty German philosophy, has never lost, amongst masses of detail, the higher philosophical standpoint. As general inspector of prisons he has certainly become acquainted with the Criminal World; and in regard to Criminal Psychology, is one of the most competent investigators of the present time.

Besides this, in prison affairs Hye has always borne the flag of humanity, and humanity has ever been the forerunner and prophet of true science.

Amidst the malicious agitation against me and my studies, I was openly supported by Hye, and the success which attended my demonstration in Paris, will doubtless be to him a satisfaction that he was among the first to discriminate mental and moral chaff from wheat.

I here also give thanks to Willhelm Pacha, student of medicine, who, with great pains, aided me in my labors.

VIENNA, SUMMER OF 1878.

PREFACE.

That man thinks, feels, desires, and acts according to the anatomical construction and physiological development of his brain, was even in olden times (*Erasistrates*) a conviction—or yet more precisely—it was a dogma among reflective natural philosophers.

The meager development of cerebral-anatomy and physiology prevented a universal dissemination of this proposition, and therefore for centuries it remained latent in the consciousness of the learned classes.

The advance of general science, the founding of craniology by Blumenbach, the interest which Gall was able to arouse by his philosophical idealism and pioneering in anatomical studies of the brain—not even yet sufficiently valued—gave a new impulse.

However greatly Gall erred in detail, the impetus he gave was very powerful, and the antitheses directed against his theses were no less productive than the theses themselves. Since then the study of the cranium and the brain has made immense strides, and scholars of all countries have helped it on, either by direct or indirect psychological investigations. I need only refer to Leuret, Gratiolet, and Broca, in France; Huschke, Virchow, and Bischoff, in Germany; Owen, Huxley and his school, in England, and Lombroso in Italy.

In spite of all contradiction in details and in special cases, the proposition of Erasistrates has received continually increasing support through the increasing knowledge of the brain and its bony cask, and every new conquest of science has been, and will continue to be, cast into that balance of the scale.

In this connection it is quite proper to ascertain whether that remarkable class of mankind, which represents the real essentials of Criminality, does not furnish data which testify in favor of the proposition mentioned.

An inability to restrain themselves from the repetition of a crime, notwithstanding a full appreciation of the superior power of the law (society), and a lack of the sentiment of wrong, though with a clear perception of it, constitute the two principal psychological characteristics of that class to which belongs more than one-half of condemned criminals.

A consideration of no less importance is the fact that the same defect of moral sensibility and will may remain concealed by superior mental organization, and greater dexterity in criminal contrivance; or it may be obscured through complications with mental disorder.

The accompanying contribution upon the cerebral constitution of criminals exhibits mainly, deficiency—deficient gyrus development—and a consequent excess of fissures, which obviously are fundamental defects. These defects are evident throughout the entire extent of the brain, and *a priori* this was to be expected, as otherwise the inclination to faulty action would have found compensation through other brain factors.

Crime is in no way analogous to monomania; it results from the psychical organization as a unit, and its particular form of expression is determined by social circumstances.

It is probable that the details of this cerebral condition, either isolated or in combination, will often be found in epileptics and in the insane, as well also as in members of encephalopathic families; the entire class will be correctly appreciated only in time to come.

Moreover, certain conditions have but a formal signification. We do not know the physiologico-psychological value of single facts.

That a defective, atypically-constructed brain, cannot function normally is so evident as to leave no ground for discussion. That which we absolutely do not know is, why such a brain acts in one certain way and not in other ways, and why it acts in just this manner under certain psychological conditions.

Another important point should be kept in view; each case should be judged of from the standpoint of race-type, and its special deviation from such type. Unfortunately, to the present, there is a lack of material for a comparative cerebro-anatomical study of races.

I hope that this publication may be a grain in the great sowing, of which the harvest shall be a true knowledge of the nature of man, and that thesis and antithesis may conduct to a lasting foundation.

From the history of Science, however, every one may derive this consolation: that no true thought and no true demonstration ever perishes, no matter how lightly they may be appreciated by contemporaneous views and feeling, or how far incomplete knowledge and defective individual talent for investigation may lead astray.'

I have tried to make these studies accessible to those not conversant with anatomy. For such, a study of the Introduction and of sections 1, 2, and 5 of the Recapitulation, will suffice.

---

[1] In reply to a question which I put to an intelligent bank-note counterfeiter, whether he would repeat the crime, he said: "*Whenever I may die, I will to you my skull and brain.*" The question of the psychology of crime seems to me to have been no more correctly answered by either Philosopher or Criminalist.

TRANSLATOR'S PREFACE.

Whatever time and labor the Translator has given to place Professor Benedikt's work before the English-reading public is regarded by him in no other light than that of a gratuitous contribution towards establishing a scientific basis for the prevention of crime.

That of course must come through a true understanding and a proper management of those born with such physical defects as entail an unusual inclination to commit crime.

The corollaries or suggestions which naturally result from Professor Benedikt's investigations lead to this end, and indicate the direction for a more rational, humane, and at the same time a more radical and secure disposition of overt criminals.

The fullest provision for public safety will be found inseparable from that course which affords also the greatest possibilities for regeneration and restored usefulness to those in whom the depraved tendency has become developed into actual habit.

Both the polity and policy of all governments have hitherto been strangely superficial and incongruous; the degree of criminality has been measured by the more or less accidental magnitude of the gross results of the criminal acts, and the public has been guaranteed temporary security only when criminal deeds have—often by the merest chance—resulted in personal or public injury and disaster.

Thus the heaviest legal penalties are often meted to those who are the least vicious, and *vice versa*.

Besides this, the entire system of past penal legislation is calculated, with the most unerring certainty, to intensify the degradation which it finds already existing with criminals, and after the government has gravely administered a legal retaliation (not correction) the subject of it is let loose upon the world, robbed of the possibility of self-respect, an irretrievable outcast, and an hundred-fold more brutish and dangerous than he was before.

That this little work may help towards bringing the more lowly organized mass of the human race up to the higher estate of noble manhood, and thus to render all classes more secure in person, property, and life; and most of all, to fit these unfortunates for the Infinite Life, is the earnest and sole desire of the Translator.

In Professor Benedikt's original works the brains of the criminals are represented by photographs. These have been reproduced by the photo-engraving process, and the Translator takes this opportunity to thank his very kind and skillful friend, M. Lorini, for giving them special personal attention. They will be found nearly, if not quite, as perfect as the original photographs, and much more plainly lettered; besides, they will not fade and become useless, as is already the case to a great degree with the photographs, though only three years issued.

38 WEST FORTIETH STREET, NEW YORK, 1880.

TABLE OF CONTENTS.

Introduction, ... 17
    1. Normal Type of Cerebral Structure, ... 17
    2. Demarcation of the Occipital Lobe, ... 23
    3. Cerebrum-Covering of the Cerebellum, . . . 26
    4. Of Confluent-Fissure Type, ... 31
Observations I—XXII, ... 33—138
Recapitulation:
    1. Statistics of Fissure-Communications, ... 139
    2. Statistics of Proportions of Cerebrum to Cerebellum, . 152
    3. External Orbital-Fissure in Man, ... 153
    4. Anthropological Law Respecting Criminals' Brains, . . 157
    5. The Application of Kant's Antinomian Doctrine to this Law, . 158
    6. Law of Development of Radiating Fissures, . . . 163
    7. Concerning the Identity of the Primate and Mammal-Brain, . 173
    8. Relation of Cerebral Conditions to Conditions of the Skull, . 182
    9. Measurements of Brain, ... 183
    10. Concerning Methods of Criminalistic-Psychology, . . 184

13

EXPLANATION OF LETTERS AND FIGURES ON THE PLATES AND WOOD CUTS.

F.—Frontal Lobes, F. 1, F. 2, F. 3,—1st, 2d, 3d, Gyri Frontales.
✦.—Upper, Secondary Gyrus of the *Gyrus Frontales Superior* in Man, and the Anterior Portion of the Upper Primary-Gyrus in Animals.
A.—Anterior Gyrus Centralis.
B.—Posterior Gyrus Centralis.
P.—Parietal Lobe: P. 1, P. 2,—1st and 2d Lobuli Parietales.
P. 2'.—Lobulus Tuberis.
O.—Occipital Lobe.
T.—Temporal Lobe : T. 1, T. 2, T. 3.—1st, 2d, and 3d Gyri Temporales.
Cbl.—Cerebellum.
Q.—Præcuneus (Lobus Quadratus.)
Cu.—Cuneus (Lobus Triangularis.)
Fu.—Gyrus Fusiformis (Gy. Occ. Temp. Lat., 4th Temp. Gyr.)
Lg.—Gyrus Lingualis (Gyr. Occ. Temp. Mc., 5th Tem. Gyr.)
H.—Gyrus Hippocampi.
U.¹—Gyrus Uncinatus (Hook-convolution.)
Gf.—Gyrus Fornicatus (Gyrus Corporis Callosi.)
Ob.—Orbital Gyrus, or Basilar Part of the Gyrus Frontalis Medius.
Ol.—Olfactory Lobe.
CC.—Corpus Callosum.
S.—Fissura Sylvii.
S'.—Fissura Sylvii, Posterior Ramus of.
S".—Fissura Sylvii, Anterior Ramus of.
f. 1, f. 2, f. 3—1st, 2d, and 3d Sulci Frontales.
φ.—Secondary Fissure of Gyrus Frontalis Superior in Man, and Anterior Portion of the Upper Primary Fissure in Animals.
c.—Sulcus Centralis (Rolando's Fissure.)

¹ In most of the cuts the U is placed too high. It is found rightly located, for example, in Figure III, Plate III. (Corrected in reproduced cuts.—*Tr.*)

(xv)

t. 1, t. 2.—Sulci Temporales, Superior and Inferior.
t. 3.—Sulcus Occipito-Temporalis (Fissure Fusiformis of Wernicke.)
ip.—Sulcus Interparietalis.
k.—Wernicke's Fissure (Fissura Occipitalis Anterior, External.)
g.—Sulcus Occipitalis Inferior.
po.—Fissura Parieto-Occipitalis (Perpendicularis.)
ho.—Sulcus Occipitalis Horizontalis.
cm.—Sulcus Calloso-Marginalis.
cc.—Fissura Calcarina.
cl.—Fissura Collateralis.
ob.—Fissura Orbitalis, or Fissura Cruciata of the under portion of the Gyri Frontales.
Of.—Fissura Olfactoria.
blt.—Fissura Basilaris Lateralis (in animals).
x.—External Orbital Fissure of Animals.
b.—Fissura Olfactoria in Animals.
h.—Scissura Hippocampi.

INTRODUCTION.

Upon the outer and upper surface of a cerebral hemisphere, as represented by Fig. 1, it is to be seen that the fissures are indicated by black lines.

In the first place there will be observed, between the frontal (F) and temporal (T) portions of the brain, a fissure which runs from before backwards and upwards, that which is called the Sylvian fissure (S) (*fissura Sylvii*). This fissure extends anteriorly by one or two branches (S″) into the frontal lobe, and posteriorly by one ascending branch (S′) into the parietal lobe (P).

Extending in a general direction from before and below, upwards and backwards, a fissure (c) courses through the middle of the brain, reaching to its upper border. A downward extension of this fissure would intersect the Sylvian fissure in the vicinity of its anterior ascending branch (S″). This fissure (c) is called the *sulcus centralis*, or *Rolando's fissure*. In typic brains it has no connection with other fissures. It divides the central portion of the outer and upper cerebral surface into two gyri, called *gyrus centralis anterior* (A) and *gyrus centralis posterior* (B).

This form of the central gyri is considered as especially characteristic of the ape and human brain. The *sulsus centralis* (c) appears in the sixth month of embryonic life.[1]

---

[1] The sylvian fissure begins to appear in the 3d month of fœtal life. The fissures of the inner surface—to be spoken of later—the *fissura calcarina* (cc) and the *parieto-occipitalis* (po) appear in the third and fourth months; the *fissura calloso-marginalis* (cm) in the fifth month.

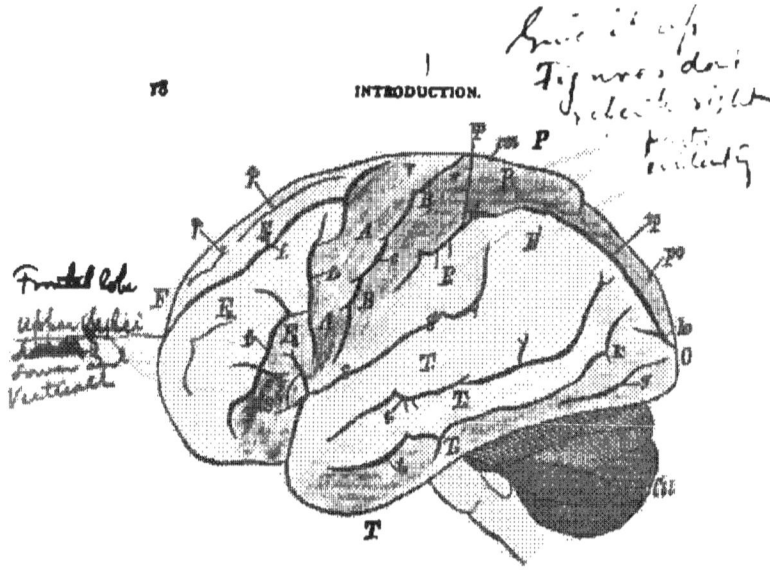

FIG. 1.
EXTERNAL SURFACE OF BRAIN.

F.=Frontal Lobe. O.=Occipital Lobe. P.=Parietal Lobe. T.=Temporal Lobe. F. 1, F. 2, F. 3.=Upper, Middle, and Lower Frontal Gyri. P. 1, P. 2, P. 3.=Upper and Anterior, and Posterior Divisions of Lower Parietal Gyri. T. 1, T. 2, T. 3.=Upper, Middle, and Lower Temporal Gyri. A. A.=Gyrus Centralis Anterior. B. B.=Gyrus Centralis Posterior. cbl.=Cerebellum. f. 1, f. 2, f. 3.=Upper, Lower, and Vertical Frontal Sulci. φ. φ.=Secondary Sulcus. v. v.=DeIle. c.=Sulcus Centralis. ip. ip.=Sulcus Interparietalis. po.=Fissura Parieto-Occipitalis. ho.=Sulcus Occipitalis Horizontalis. S.=Fissura Sylvii. S'.=Ramus Fissurae Sylvii Posterior. S".=Rami Fissurae Sylvii Anterior. t. 1, t. 2.=Upper and Middle Temporal Sulci. k.=Wernicke's Sulcus Occipitalis Anterior. g.=Sulcus Occipitalis Inferior.

In the angle between the anterior ascending branch (S") of the sylvian fissure and the *fissura Sylvii* (S) is to be seen a second radiating fissure (f. 3), which, to a certain extent, runs parallel with the *sulcus centralis*.

This is the so-called third frontal fissure (f. 3) (*sulcus frontalis perpendicularis*). In typical brains this fissure has no connection with the sylvian fissure, though it often has (as shown by dots in the drawing) with the lower frontal fissure (f. 2) (*sulcus frontalis inferior*).

INTRODUCTION. 19

The third frontal fissure is also called the *præcentral fissure*. I reserve this term for another use, that is, in event of the third frontal fissure running nearly parallel with the *sulcus centralis* along the greater part of the external surface, nearly up to the superior medial border. (See Fig. A, p. 15). It then blends with the vertical branch which is to be seen (Fig. 1) coming from the *sulcus frontalis superior* (f. 1), and usually also with the vertical branch of that fissure, which generally forms a Y-shaped depression in the upper portion of the *gyrus centralis anterior* (A). (See Fig. 1 and Fig. A, of Introduction, also Fig. I, table I, etc.). An extreme development of such a præcentral sulcus naturally indicates a dwarfage, especially of the anterior central gyrus, through an unusual demand for fissure-space.

In the space between the *sulcus centralis* (c) and the posterior ascending branch of the sylvian fissure (S′) is to be seen another radial fissure (ip), which extends, sometimes divided into two parts and sometimes uninterruptedly, to the occipital lobe (O, Fig. I). It is the so-called inter-parietal fissure (ip) (*sulcus interparietalis*).

This fissure is also called the retrocentral fissure (*retrocentralis*). Again I also reserve this term, as in the case of the præcentral fissure, for those instances where the radial portion of the *sulcus interparietalis* ascends parallel with the *sulcus centralis* to the medial border. (See for example, Fig. II, Plate xii, and Fig. A, p. 29.) This construction arises from the blending with the *sulcus interparietalis* of the fissures coming from within and around the upper third of the posterior central gyrus.

Besides these three important radial fissures, which to a certain extent run parallel with each other, there is another set of fissures (*sagittal*) directed antero-posteriorly.

There are two of these in the frontal portion (F) of the brain. The first (f. 1) and the second (f. 2) frontal fissures. The first separates the upper frontal gyrus (F. 1) from the middle frontal gyrus (F. 2). The second frontal fissure (*sulcus frontalis inferior*) (f. 2) separates the middle frontal gyrus (F. 2) from the lower (F. 3). We may here specially mention that the generally shallow fissure in the upper frontal gyrus (F. 1), indicated by φ, sometimes become very deep and separates the upper frontal gyrus (F. 1) into two gyri.

The *sulcus interparietalis* (ip) separates the parietal portion of the brain (P) which lies back of the posterior central gyrus (B) into two parts. That above the *sulcus interparietalis* (ip) is the first parietal lobule (P. 1). That which lies beneath and partly limited on its lower edge by the *fissura Sylvii* (S) is the second parietal lobule (P. 2). The posterior portion of this lobule is generally regarded as an independent lobule (P. 2′), and is called also *Lobulus tuberis* (f) (*gyrus angularis*).

The temporal lobe (T) contains two main fissures which, in the typic brain, run more or less parallel with the *fissura Sylvii*. They are the upper (t. 1) and lower (t. 2) *sulci temporales*. The last is often separated into two parts. Both may extend very high up into the vicinity of the *sulcus interparietalis*.

A connection of the inferior *sulcus temporalis* (t. 2) with the *interparietalis* is considered as atypic.

The gyrus between the sylvian fissure (S and S′) and the upper temporal sulcus (t. 1) is known as the upper (first) temporal gyrus (T). The space between the upper and lower temporal sulci (t. 1 and t. 2) lodges the second temporal gyrus (T. 2) (*gyrus temporalis medius*). Beneath the second temporal sulcus (t. 2) (*sulcus temporalis medius*) lies the third (inferior) temporal gyrus (T. 3).[1]

On the upper border at the posterior limit of the parietal lobe (P) may be seen a fissure which extends from the inner surface to the upper external surface of the hemisphere (po). This is the parieto-occipital (perpendicular) fissure (po). It separates the parietal (P) from the occipital (O) lobe. In the

---

[1] In case no third temporal sulcus exists (t. 3 of Fig. 1) there is also no distinct third temporal gyrus. Then the gyrus of the base, which stands next in succession—the *gyrus Hippocampi* or the *gyrus fusiformis*, appears artificially widened. *Wernicke's fissura fusiformis* represents sometimes the second and sometimes the third temporal sulcus, as termed by authors. Further along we shall recur to the significance of this fissure.

occipital lobe there is of most special note the sulcus occipitalis horizontalis (ho).

A second fissure that extends from the inner surface of each hemisphere to its upper and outer surface is the so-called sulcus calloso-marginalis (cm). It is of topographical importance, because it defines the separation between the central lobe (B) and the parietal lobe (P), and in atypic brains is a guiding point of great assistance. A fissure which I call, in honor to its discoverer, *Wernicke's fissure* (k), serves as a line of demarcation on the one side between the parietal and temporal lobes (P and T), and on the other the occipital lobe (O). The second temporal gyrus (T. 2) is always situated in front of Wernicke's fissure (k). It lies in an imaginary arc situated between the parieto-occipital fissure (po) above, and the second temporal sulcus (t. 2) below. These three fissures (po, k, and t. 2) combined, form, in apes, the so-called "*ape-fissure*," which divides the temporal and parietal lobes (T and P) from the occipital (O).[1]

II.

Figure 2 represents the inner cerebral surface as it appears after a division of the commissures which anatomically connect the hemispheres. To these commissures belongs especially the *corpus collosum*, the medial longitudinal section of which is seen at CC in Fig. 2.

Nearly parallel with the corpus callosum (CC) runs a sulcus which bears the name of *sulcus calloso-marginalis* (cm). It ascends behind the posterior central gyrus (D) to the upper hemispheric border, and thus separates the so-called central lobe (of Betz) from the parietal division of the brain (P).

The gyrus between the *fissura calloso-marginalis* (cm) and the *corpus callosum* (CC) is termed *gyrus fornicatus* (Gf), or the inner border-lobe (*gyrus corporis callosi*). This gyrus runs in the same direction as, and further posteriorly than the *sulcus calloso-marginalis*.

---

[1] The fissure (g.), called by Wernicke *fissura occipitalis inferior*, separates the occipital lobe of the external surface from the base.

22   INTRODUCTION.

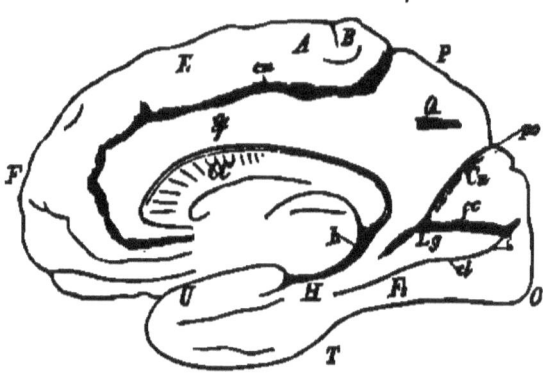

FIG. 2.
(MEDIAN SURFACE OF BRAIN.)

F.=Frontal Lobe.   O.=Occipital Lobe.   P.=Parietal Lobe.   T.=Temporal Lobe.   F. 1.=Gyrus Frontalis Superior.   A.=Gyrus Centralis Anterior.   B.= Gyrus Centralis Posterior.   Q.=Præcuneus.   Cu.=Cuneus.   Lg.=Gyrus Lingualis.   Fu.=Gyrus Fusiformis.   U.=Gyrus Uncinatus.   H.=Gyrus Hippocampi.   Gf.=Gyrus Fornicatus.   cm.=Sulcus Calloso-Marginalis.   po.=Fissura Parieto-Occipitalis.   cc.=Fissura Calcarina.   cl.=Sulcus Collateralis.   h.= Scissura Hippocampi.   CC.=Corpus Callosum.

The strongly developed *parieto-occipitalis* (po). is now observed. From above and behind, it extends downwards and forwards, and in the human brain always blends with another fissure which runs more horizontally, that called the *fissura calcarina* (cc).

The *fissura parieto-occipitalis* (po), with the *fissura calcarina* (cc), makes a fork, and as concerns the human brain, it seems a matter of indifference to which fissure the handle of this fork is assigned.

The gyrus (F. 1, Fig. 2) which, in the anterior portion of the medial cerebral surface, runs parallel with the *gyrus fornicatus* (Gf,) is considered as belonging to the upper frontal gyrus (F). The posterior limit of this gyrus (F) on the medial surface is not, as a rule, sharply defined, and therefore it merges into the central gyrus (A). Often there is a distinct separation, and then the medial part of the central lobe, as the "para-central lobe" of Betz, is well defined (See Fig. B). I will here say incidentally that I saw such a demarcation of a "para-central lobe" in a bear's brain. The *sulcus centralis* (c) generally extends over the upper medial border into the medial surface, thus sharply dividing on the medial surface the anterior central gyrus (A) from the posterior central gyrus (B).

That portion of the medial parietal lobe (P) which lies between the ascending branch of the *sulcus calloso-marginalis* (cm) and the *fissura parieto-occipitalis* (po) is the *præcuneus* (Q), whilst the gyrus (cu) which lies in the fork made by the *fissura parieto-occipitalis* (po) and the *fissura calcarina* (cc) is called the *Cuneus*.

The demarcation between the posterior arch of the *gyrus fornicatus* (Gf) and the *præcuneus* is seldom thoroughly defined.

At its base, in the posterior fossa of the cranium, the brain slopes not only downwards but also inwards, from which it results that this portion of the brain can be seen from an inside view. Running parallel with the *fissura calcarina* (cc) is to be seen another important fissure; the *sulcus collateralis* (cl), which divides that portion of the base of the brain lying beneath the *fissura calcarina* (cc); the upper division being called the *gyrus lingualis* (Lg); the lower part, the *gyrus fusiformis* (Fs).

From the internal view can also be seen the lobe of the middle cranial fossa, that is, the so-called hook-gyrus or *gyrus uncinatus* (U) and the *gyrus hippocampi* (H).

Above the *gyrus hippocampi* (H) there is a deep fissure termed *scissura hippocampi* (h).

III.—BASE.
Fig. 3.

The base of the brain includes three sharply-defined divisions, the boundaries of which correspond to the raised osseous ridges between the three cranial fossæ.

To the most anterior portion belongs the frontal lobe (F), in which we first observe a fissure running parallel with the inner border, the *sulcus olfactorius* (of); between this sulcus and the inner border lies the olfactory lobe (Of) (*gyrus rectus*). Externally to this there lies a second gyrus, which I will term the orbital gyrus (or lobe) (Ob). (*Gyrus fron-*

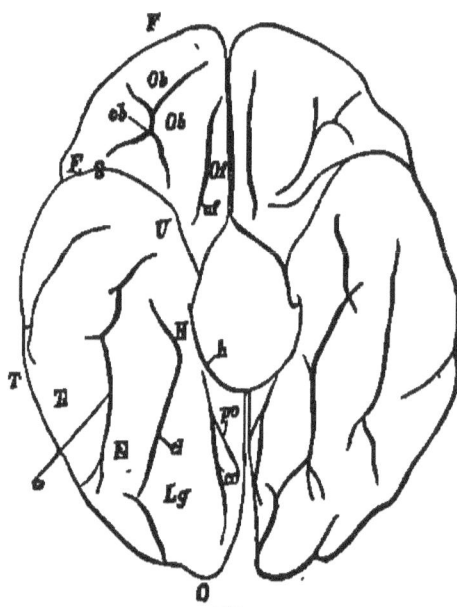

FIG. 3.
(BASE OF BRAIN.)

F.=Frontal Lobe. O.=Occipital Lobe. T.=Temporal Lobe. F. 3.=Gyrus Frontalis Inferior. Ob Ob.=Sulcus Orbitalis. U.=Gyrus Uncinatus. H.= Gyrus Hippocampi. Lg.=Gyrus Lingualis. Fs.=Gyrus Fusiformis. T. 3.= Gyrus Temporalis Inferior. ob.=Sulcus Orbitalis. of.=Sulcus Olfactorius. S.=Fissura Sylvii. h.=Scissura Hippocampi. po.=Fissura Parieto-Occipitalis. cc.=Fissura Calcarina. cl.=Sulcus Collateralis. t 3.=Sulcus Temporalis Inferior.

talis medius) (*base*). It contains an H-formed sulcus, which is called orbital sulcus (ob). [1]

At the external border there is to be seen a bent-over piece of the *gyrus frontalis inferior* (F. 3).

The anterior division (F), which has just been described, is separated from the middle division by a cleft, which connects with the *fissura sylvii* (S).

This middle part, the "middle basilar lobe," as we have just learned from the medial representation, is composed of the *gyri uncinatus* (U) and *hippocampi* (H). (See Fig. 2.) This middle portion of the base is separated from the posterior basilar portion, which lies in the posterior cranial fossa, and which is composed of the *gyri lingualis* (Lg) and *fusiformis* (Fs). The separation is not effected by a transverse fissure, but by a more or less sharp edge which runs from within and forwards, pursuing a direction outwards and backwards. [2]

It is only in the atypic brain, as will be seen in the following section, that there extends from the *fissura collateralis* (cl) or the *fissura fusiformis*, a transverse fissure which more or less sharply and completely separates the middle from the posterior basal lobe.

That part of the cerebral base which lies back of the ridge, leaves the level of the *gyrus uncinatus* (U) at that point, and ascends to the natural position of the brain back and above. [3]

The *gyrus temporalis inferior* (T. 3) is found at the base in the region of the middle cranial fossa; it seldom extends back into the posterior cranial fossa. This third temporal gyrus (T. 3) has, however, no very constant formation.

---

[1] This orbital lobe is generally reckoned as a part of the *gyrus frontalis medius* (F. 2), but for potent reasons, derived from comparative cerebral-anatomy, there should be assigned to this gyrus a greater degree of independence from the external frontal lobe. Even anteriorly, it is frequently separated from the middle and lower (external) frontal gyri by a fissure described by Wernicke, as the *fissura fronto-marginalis*.

[2] In the drawing the *gyri lingualis* and *fusiformis* appear on the same level with the *gyri uncinatus* and *hippocampi*. In reality, however, in a view from below, the first-named gyri lie much deeper than those last named.

[3] As the *fissura collateralis* (cl of Figs. 2 and 3) is absent in many animals, in these cases the blended *gyri lingualis* and *fusiformis* represents a fused basilar occipital lobe. We shall revert to this.

IV.

There is the greatest disagreement among authors respecting the boundary of the occipital lobes. According to Ecker the *gyri lingualis* and *fusiformis* (Lg and Fs in Figs. 2 and 3) belong to the temporal lobe. This is certainly unwarrantable, for all of the cerebrum which is contained in the posterior cranial fossa, and all that back of the *splenium corporis callosi* (the posterior end of the corpus callosum, CC, Fig. 3) should be considered as belonging to the occipital lobe. Here commences the region of the posterior continuation of the *cornu ammonis*, in the interior of the brain, and that is to be recognized as the ganglion of the occipital lobe, which, upon the external surface, has the *fissura calcarina* (cc) as its companion.

The ridge, before described, between the *gyrus uncinatus* (U) on the one side and the *gyri lingualis* and *fusiformis* on the other, constitutes the border of the base; the *fissura parieto-occipitalis* (po) and the common continuation of this and the *fissura calcarina* (cl) constitute the boundary on the medial surface.

When Wernicke's anterior occipital fissure (k, Fig. 1) does not exist on the surface, an imaginary curved line must be projected from the *fissura parieto-occipitalis* (po), parallel with the *sulcus temporalis superior* (t. 1) and through the *sulcus temporalis medius* (t. 2) to the lower external border. This would be the boundary, on the external surface, of the occipital lobe. It corresponds to the idea of the occipital lobe in the ape's brain.[1]

V.

The proportion of the cerebellum (cbl, Fig. 1) to the cerebrum is also very important. Upon the removal of the European brain from the scull, the cerebellum is found completely covered by the cerebrum. According to Retzius, the brain of the Laplander affords the only exception. The same author

---

[1] Otherwise expressed, it may be said that a plane which intersects the junction of the lamboidal and sagittal sutures on the one hand, extending on the other hand through the upper border of the pyramid, will separate, upon each hemisphere, the occipital lobe from the temporal and parietal lobes.

says also, that in the Teutonic and Latin races the cerebellum is very thoroughly covered by the occipital lobes, whilst among the Sclavonians it is barely covered.

I know of no positive information concerning the proportions as they exist in the Finnic-Magyaric or in the Gypsy races. From my experience, however, it would appear to be much the same as with the Sclavonic race. Respecting people outside of Europe there is no positive knowledge.

Upon the brain of a "Charruas," in the Atlas of Gratiolet and Leuret, the cerebellum is only barely covered.

It becomes self-evident that a relative shortening of the occipital lobes is a matter of importance when we consider that in the inferior types of apes, and also in the entire remaining range of animals, these lobes (O) are not sufficient to cover the cerebellum. As a rule, microcephala exhibit the same deficiency, although the covering is found to be present even in human fœtus.

When in the Teutonic or Latin races this covering is found to be scanty, we may regard it as an insufficiency of the occipital lobes, at the same time it must be observed that the shortening of the cerebral hemispheres need not pertain exclusively to the occipital lobes; it may result also from a dwarfage of the anterior or middle lobes.

VI.

In opposition to the normal type, which we have up to this point been describing, another type is presented in the accompanying description of brain-specimens from criminals. (See Figs. A and B.)

The most important characteristic of this type consists in this: *If we imagine the fissures to be water-courses, it might be said that a body floating in any one of them could enter almost all the others.* There are also absent a great number of annectants, which are important cerebral territories, the absence of which represents so many aplasias.

For some time a marked fissuring of the brain was regarded, erroneously however, as a sign of high development. It is true that if, in the ascending scale of animal life, there appears a new typical fissure, it signifies, as a rule, an extended development of the surrounding cerebral region. But where there is no new development around the fissure, and especially where the more marked fissure results from a junction of typical fissures, the fissure thus emphasized indicates a defect arising from the absence of annectants.

If we observe the fissures in Fig. A, p. 29, which represent the collective fissure formation in the brains which we shall describe, it is primarily observable that the three important radiating fissures upon the cerebral surface, to wit: the *sulcus centralis* (c) (at 2, Fig. A), the third frontal fissure (f. 3) (at 3, Fig. A) and the radial portion of the *sulcus interparietalis* (ip) (at 4, Fig. A), all show a great inclination to unite with the *fissura Sylvii* (S), so that we now have not only an anterior and posterior ascending branch of the *fissura Sylvii*, but also three other radiating branches, namely: c, f. 3, and ip.

As the two latter fissures have, moreover, an inclination to extend themselves vertically upwards to the medial border (7 and 13, Fig. A), it results that there are frequently three central parallel fissures, the third frontal fissure (f. 3) (as before mentioned) appearing as a *præcentralis* and the *sulcus interparietalis* (ip) as a *retrocentralis*. This formation of the præcentral and retrocentral fissures is not accomplished by prolongations, but through blendings, and this indeed in a measure with fissures that in the typic brain are scarely indicated.

In such brains the determining of one's whereabouts would sometimes be very difficult if the *sulcus calloso-marginalis* (cm, Fig. A) were not a secure guide. The ascending portion of this sulcus (see Fig. B) forms the posterior border of the *gyrus centralis posterior* (B) and the first radiating fissure in front of this (on the outer surface) *must* be the *sulcus centralis* (c).

In the frontal lobe, one or the other of the frontal fissures (f. 3, f. 2, f. 1, or φ) is often connected with the *sulcus centralis* (c). (See for example 5 in Fig. A.) On some of these brains the secondary fissure (φ Fig. I) is deep and long; in which case it often penetrates deeply into the upper part of the *gyrus centralis anterior* (A), and participates in the con-

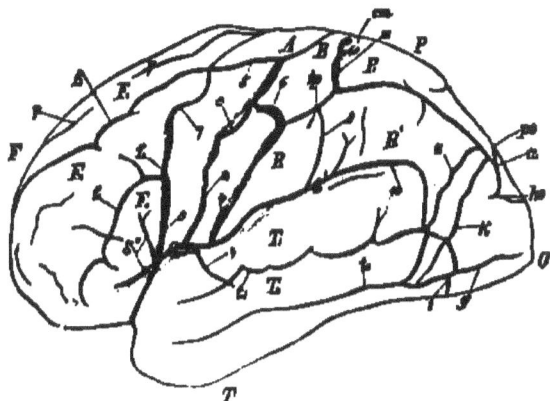

FIG. A.
EXTERNAL SURFACE OF BRAIN FISSURE-TYPE.

F.=Frontal Lobe. O.=Occipital Lobe. P.=Parietal Lobe. T.=Temporal Lobe. F. 1, F. 2, F. 3.=Upper, Middle, and Lower Frontal Gyri. P. 1, P 2, P. 2′.=Upper and Anterior, and Divisions of Lower Parietal Gyri. T. 1, T. 2, T. 3.=Upper, Middle, and Lower Temporal Gyri. A. A.=Gyrus Centralis Anterior. B. B.=Gyrus Centralis Posterior. clb.=Cerebellum, f. 1, f. 2, f. 3.= Upper, Lower, and Vertical Frontal Sulci. ϑ. ϑ.=Secondary Sulcus. v. v.= Delta. c.=Sulcus Centralis. ip. ip.=Sulcus Interparietalis. pa.=Fissura Parieto-Occipitalis. ho.=Sulcus Occipitalis Horizontalis. S.=Fissura Sylvii. S′.=Ramus Fissuræ Sylvii Posterior. S″.=Rami Fissuræ Sylvii Anterior. t. 1, t. 2.=Upper and Middle Temporal Sulci. k.=Wernicke's Sulcus Occipitalis Anterior. g.=Sulcus Occipitalis Inferior. 1.=Connection of Upper and Middle Temporal Sulci with Inter-Collateralis. 2.=Connection of Sulcus Centralis with Fissura Sylvii. 3.=Connection of 3d Sulcus Frontalis with Fissura Sylvii. 4.= Connection of Sulcus Interparietalis with Fissura Sylvii. 5.=Connection of Frontal Sulci with Sulcus Centralis. 6.=Connection of Sulcus Interparietalis with Sulcus Centralis. 7.=Extension to Medial Border of 3d Frontal Sulcus. 8.=Connection of Interparietalis with Ramus Posterior Fissuræ Sylvii. 9.= Connection of Sulcus Tempor. Sup. with Fissura Sylvii. 10.=Connection of Upper and Middle Sulci Temporales with Fissura Sylvii. 11.=Connection of Upper and Middle Sulci Temporales with Occipital Fissures. 12.=Connection of Parieto-occipitalis with Occipitalis horizontalis. 13.=Extension of Interparietalis to Medial Border.

struction of the præcentral fissure. Under this circumstance we have really four frontal gyri as found in beasts of prey, such as the cat and fox.

In the parietal lobe (P) we often see the *sulcus interparietalis* (ip.) rising out of the *fissura Sylvii* (S) (4 in Fig. A); moreover, it not infrequently blends with the posterior ascending ramus (S') of the *fissura Sylvii* (as at 8, Fig. A). Also there often exists a transverse connection with the *sulcus centralis* (c) (6, Fig. A). And, furthermore, the *interparietalis* (ip.) unites with the upper or middle temporal fissures (t. 1 and t. 2) (10, Fig. A). As the *interparietalis* (ip.) generally joins with the *occipitalis horizontalis* (ho) and this last, in the brains which we will describe, often connects with the *parieto-occipitalis* (po) (12, Fig. A), a connection is thereby effected between the *interparietalis*, the medial surface, and the *fissura calcarina* (cc).

In the temporal lobe (T) we often see the upper temporal sulcus (t. 1) in connection with the *fissura Sylvii* (S) by means of a transverse fissure (9, Fig. A) or it may, as before mentioned, connect with the *interparietalis* (ip. 10) or with the occipital fissures (11, Fig. A). The first and second temporal fissures (t. 1, t. 2) also frequently communicate with each other (1, Figs. A and B) and send out a connection around the outer and under border of the brain, to the *fissura collateralis* (cl).

Again, the lower occipital sulcus (g) often unites with the *fissura fusiformis* (t. 2 or t. 3 of authors) and as it (g) often springs from the upper temporal sulcus (t. 1) and the last one (*fiss. fusif.*) not infrequently unites with the *sulc. collateralis* (cl) and this in its turn with the *fissura calcarina* (cc). So in this way there results in these brains an extensive connection between the temporal and occipital basilar lobes. On the medial surface (compare Figs. 2 and B) we see that the *sulcus calloso-marginalis* (cm) sends a prolongation, more rarely to the *parieto-occipital fissure* (po), but often to the common stem of this (po) and the *fissura calcarina* (cc).[1]

In some brains the *sulcus centralis* (c) which, in these cases,

---

[1] This connection is effected by penetrating the fissures of the præcuneus (Q), which has been insufficiently indicated in Fig. 2.

penetrates far into the medial surface, is also connected with the *sulcus calloso-marginalis* (cm).

Morever, we see the *fissura calcarina* (cc) united in an atypic manner with the *scissura hippocampi* (h).

The *fissura calcarina* (cc) often communicates also with the *collateralis* (cl).

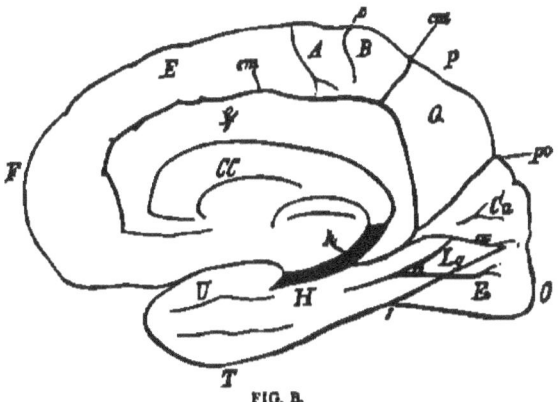

FIG. B.

MEDIAL SURFACE OF BRAIN FISSURE-TYPE.

F.=Frontal Lobe. O.=Occipital Lobe. P.=Parietal Lobe. T.=Temporal Lobe. F. 1.=Gyrus Frontalis Superior. A.=Gyrus Centralis Anterior. B.= Gyrus Centralis Posterior. Q.=Praecuneus. Cu.=Cuneus. Lg.=Gyrus Lingualis. Fs.=Gyrus Fusiformis. U.=Gyrus Uncinatus. H.=Gyrus Hippocampi. Gf.=Gyrus Fornicatus. cm.=Sulcus Calloso-Marginalis. po.=Fissura Parieto-Occipitalis. cc.=Fissura Calcarina. cl.=Sulcus Collateralis. h.= Sciasura Hippocampi. CC.=Corpus Callosum. 1.=Connection of Upper and Middle Temporal Sulci with Sulcus Collateralis.

Between the normal type with the typically-separated fissures, and the confluent-fissure type, there exist manifold transitions, inasmuch as more or less of the annectant gyri are not developed at the surface, but remain concealed as underlying annectant gyri (*Plis de passage*, in the stricter sense).

In a former series of contributions respecting the brains of criminals, I have emphasized the absence of annectants. In this I have especially shown an important anatomical fact, that is, a deviation from the school-type of brain. This statement found confirmation, and the opportunity was used, in a curious ethical discussion, to rouse suspicion against me. In connection with this it must be kept in mind that it is not a single deviation by itself which constitutes the characteristic of the type, but rather a general deviation. Those brains which exhibit numerous deviations from the first type and approach the second, we shall rank under the head of the second, and brains with few or isolated deviations from the first type, we shall consider as belonging to that type.

There is no doubt whatever that this second type is developed in embryo the same as it is found in the mature adult brain. Even Rüdinger[1] has shown that in the foetal stage of existence the brachio and dolicho-cephalic brains exhibit their characteristic differences.

Most important conclusions would be derived from a comparative race-study of brains if it should be demonstrated that in the lower human races the brain entire or in various parts corresponds more or less to the second type. If we observe the negro brain, in the celebrated work of Huschke's (Table vi, Fig. III), we see that it belongs decidedly to the second type.

Before proceeding further it must be observed that the brains of inferior individuals most certainly approach, as a rule, nearer to the second than the first type.

If we consider what kind of material usually finds its way into dissecting-rooms we should naturally expect to meet there a larger proportion of brains which approach the second type; and such is the general fact. The principal material of the dissecting-room consists of the remains of those who have suffered complete shipwreck in life through a low grade of intelligence, imperfect motor development, or through crimes and vice; for instance: inebriates, epileptics, prostitutes, etc. Neither by themselves or through their social connections could they procure or save the means necessary for their burial.[2]

---

[1] Unterschiede der Grosshirnwindungen nach dem Geschlechte beim Fötus und Neugeborenen, etc. (München, Riedel, 1877).

[2] The same observation applies to skulls of anatomical collections.

OBSERVATION L

(TABLE I.)

BALÁZS, a Roumanian, middle aged; as a convict good-natured, diligent, and exemplary. He was a thief and a receiver of stolen goods, and was condemned for participation in robbery, accompanied by murder. The accomplices declared that he knew of the murder, that he had kept it secret, but that he had not participated in it.

The cerebellum was incompletely covered by the occipital lobes of the cerebrum.

LEFT HEMISPHERE (SEE TABLE L FIGS. I-II).

The *sulcus centralis* (c, Fig. II) unites with the *fissura Sylvii* (S); and by two branches (3, 3, Fig. I) communicates also by one, with the *sulcus frontalis superior* (f. 1), *sulcus frontalis perpendicularis* (f. 3), and by the other with the *sulcus interparietalis* (ip). The *sulcus frontalis perpendicularis* (f. 3) runs parallel with the *sulcus centralis* (c), and receives the vertical branch of the *sulcus front. super.* (f. 1, Fig. I, Table I), thus forming a præcentral fissure.

Relatively, the *gyrus frontalis superior* (F. 1) is poorly developed.

The *sulcus interparietalis* (ip) is imperfectly separated from the *fissura Sylvii* (S), and is divided into two parts: the posterior part connects with the *fissura occipitalis horizontalis* (ho, Fig. II). The anterior portion sends downwards (at 1, Fig. II) a branch which penetrates the lower parietal lobe (P. 2 and P. 2') and communicates with the *sulcus temporalis*

OBSERVATION I.

Balázs—Murderer and Robber.

(Roumanian.)

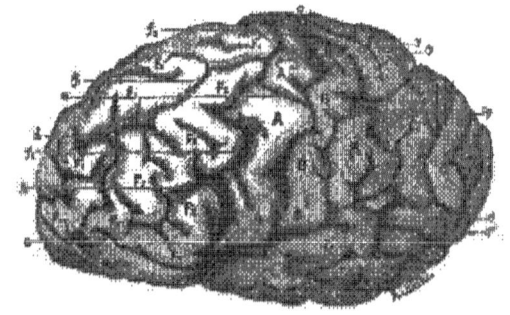

I.

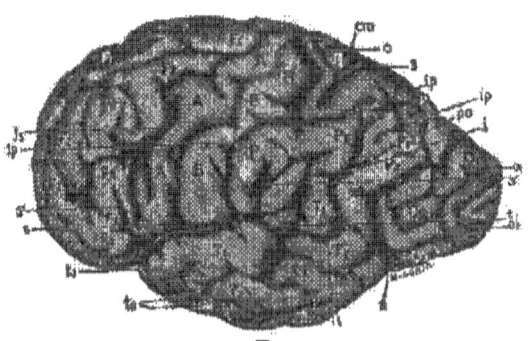

II.

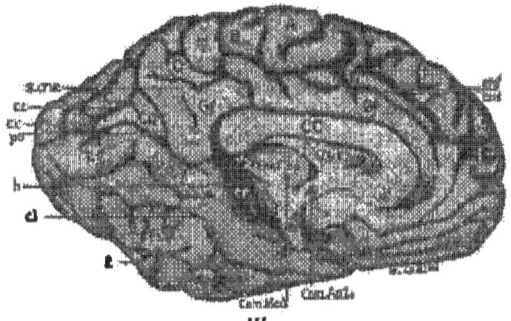

III.

A.—Gyrus Centralis Anterior.
B.—Gyrus Centralis Posterior.
Cu.—Cuneus.
CC.—Corpus Callosum.
F. 1.—Gyrus Frontalis Superior.
F. 2.—Gyrus Frontalis Medius.
F. 3.—Gyrus Frontalis Inferior.
Fu.—Gyrus Fusiformis.
Gf.—Gyrus Fornicatus.
H.—Gyrus Hippocampi.
Lg.—Gyrus Lingualis.
P. 1.—Lobulus Parietalis Superior.
P. 2.—Lobulus Parietalis Inferior.
P. 2′.—Gyrus Angularis (Lobulus Tuberis-Benedikt).
Q.—Lobulus Quadratus (Præcuneus).
S.—Fissura Sylvii.
S′.—Fissura Sylvii, Posterior Ramus.
S″.—Fissura Sylvii, Anterior Ramus.
T. 1.—Gyrus Temporalis Superior.
T. 2.—Gyrus Temporalis Medius.
T. 3.—Gyrus Temporalis Inferior.
U.—Gyrus Uncinatus.
O.—Occiput.
c.—Sulcus Centralis.
cc.—Fissura Calcarina.
d.—Sulcus Collateralis.
cm.—Sulcus Calloso-Marginalis.
f. 1.—Sulcus Frontalis Superior.
f. 2.—Sulcus Frontalis Inferior.
f. 3.—Sulcus Frontalis Perpendicularis.
h.—Scissura Hippocampi.
ho.—Fissura Horizontalis.
ip.—Sulcus Interparietalis.
po.—Fissura Parieto-Occipitalis.
s. croc.—Sulcus Cruciatus.
t. 1.—Sulcus Temporalis Superior.
t. 2.—Sulcus Temporalis Inferior.
q.—Secondary Sulcus Frontalis.

superior (t. 1) and indirectly with the *ramus ascendens* (S′) of the *fissura Sylvii*.

The *sulcus tempor. supr.* (t. 1) gives off two upper branches, one of which (1, Fig. II) blends with the *interparietalis* (ip), the other including with it *Wernicke's fissure* (k), unites with the *parieto-occipitalis* (po). There exists also a shallow union with the *fissura Sylvii*. The two occipital fissures (po and ho) are also joined together by a shallow fissure.

At 2, a fissure (Fig. II and III) leaves the *temporalis superior* (t. 1), runs backwards and downwards to the base of the brain, makes a fork-like separation, the prongs of which embrace the *gyrus fusiformis* (Fs) (*gyrus occipito-temporalis lateralis*), and proceed to communicate with the *fissura collateralis* (cl).

The *gyrus fusiformis* (Fs) exhibits frequent transverse divisions; the *gyrus lingualis* (Lg) (*gyrus occipito-temporalis medius*) very narrow. The *calloso-marginalis* (cm) connects with the fissures of the *præcuneus*, and they conjointly extend nearly to the *parieto-occipitalis* (po).

The central and posterior base-lobes are separated from each other by transverse fissures coming from the *collateralis* (cl).

Antero-posterior chord,[1]	16.7
Hemispheric arch,	24.2
Anterior curve,	13.7
Middle curve,	5.5
Posterior curve,	5.0

RIGHT HEMISPHERE.

The *sulcus centralis* (c) connects with the *f. Sylvii* (S) and the *præcentralis*. The two *gyri centrales* (A and B) moderate-sized in the lower third, are poorly developed in the upper third.

---

[1] "Antero-posterior chord" signifies the chord between the extreme frontal and occipital points. It corresponds to the longitudinal arc-chord on the skull. The curve between the two points is the "*Hemispheric arch*." It corresponds approximatively to the longitudinal arch of the skull. This arch is divided into three portions: the frontal or "anterior" extends from the before-mentioned frontal extremity to that point where the *sulcus centralis*, or an imaginary contin-

The third frontal fissure (f. 3) is imperfectly separated from the *fossa Sylvii*, and with the vertical branches of the *s. frontalis superior* (f. 1); joined by a secondary fissure (✱), it forms a well-developed *præcentralis*.

The *s. frontalis superior* (f. 1) is composed of two pieces, the anterior one of which furnishes the vertical limb of the *præcentralis*. A posterior part of the secondary fissure (✱), the same as the *f. front. sup.* (f. 1), gives still another radiating branch.

The *gyrus frontalis superior* (F. 1) contains three deep, secondary fissures, parallel with f. 1, the foremost of which sends a vertical branch contributing to the formation of the *præcentralis*.

The *s. interparietalis* (ip) is divided into an anterior and a posterior part. The first portion forms a well-developed retrocentral fissure, and by an inferiorly directed branch communicates with the *s. temporalis superior* (t. 1); it is also poorly separated from the posterior ascending branch of the *fissura Sylvii* (S′).

The *parieto-occipitalis* (po), by a very shallow anteriorly extending branch, communicates with the posterior portion of the *s. interparietalis* (ip) and also with the *occipitalis horisontalis* (ho).

(On account of an injury to the specimen, it could not be determined if the *s. temporalis superior* communicated with the *occipitalis horisontalis* (ho).)

The *f. parieto-occipitalis* (po) communicates with the *scissura hippocampi* (h). The *gyrus lingualis*, both longitudinally and transversely, is frequently divided. The middle

---

nation of the same would strike the medial hemispheric border; it is not identical with the *frontal curve* of the skull. From the last designated point, the parietal or "middle" arch extends to that point where the *parieto-occipitalis* (po) intersects the medial hemispheric arch; it does not correspond to the *parietal curve* of the skull, but the "anterior" and "middle" portions combined correspond nearly to the totality of the frontal and parietal curves on the skull. The occipital or "*posterior*" curve extends from the posterior termination of the "middle" curve to the occipital apex. Further along it will be explained that the last curve has no real correspondence with the curve between the posterior fontanelle and the *prominentia occipitalis maxima;* the value which these measurements generally have will also be explained.

basilar lobe is separated from the posterior one by a transverse fissure. The *s. calloso-marginalis* (cm) communicates with the fissures of the *præcuneus* (Q).

Antero-posterior chord,	. . . .	15.8
Hemispheric arch,	. . . . .	24.0
Anterior curve,	. . . .	13.5
Middle curve,	. . . .	4.7
Posterior curve,	. . . .	5.8

APPENDIX TO OBSERVATION L.

RIGHT: The *parieto-occipitalis* (po) communicates with the *s. cruciatus* of the *præcuneus* (Q). There is no communication between the *calloso-marginalis* (cm) and the fissures of the *præcuneus* (Q).

The external orbital sulcus is composed of two parts, and is very deep. The posterior part separates extensively the frontal lobe (F) from the orbital lobe (Ob), but does not communicate with the third incision of the *fossa Sylvii* (S).

LEFT: The anterior and middle parts of the external orbital sulcus (ob) communicate with each other and with the *fossa Sylvii* (S), but not with the third incision of the *fossa Sylvii*, which appears as a fourth incision.

OBSERVATION II.

(TABLE II.)

MADÁRASZ, János, aged 43, Selavonian, habitual thief, was finally condemned for burglary; twice escaped from prison. Of a sweet, fawning behavior in prison; treacherous and cowardly toward his overseers.

The right cerebellum imperfectly covered. The upper surface of the cerebellar hemispheres slope abruptly off, and the *vermiformis* is crowded wedge-shaped, in between the hemispheres. The posterior and medial parts of the cerebellum are uncovered, and on the left, the middle part is also uncovered.

RIGHT HEMISPHERE (TABLE II, FIGS. I–III).

The *s. centralis* (c) is well separated from the *f. Sylvii* (S). The *s. front. perpend.* (f. 3) unites with the *fossa Sylvii* (S). The *s. frontalis superior* (f. 1) by means of a descending branch communicates with the *s. frontalis inferior* (f. 2) and thereby with the *s. front. perpend.* (f. 3), and also with the *f. Sylvii* (S). The *s. interparietalis* (ip) also (Figs. I and II) communicates with the *f. Sylvii* (S and S'), and on the medial surface, by means of the *s. cruciatus*, makes an indirect and shallow connection with the *f. parieto-occipitalis* (po). In Fig. II can trace *interparietalis* (ip) to the upper medial border. In Fig. III, the fissure, entering from the medial border between cm and po, is the *s. interparietalis*. The *ramus posterior fissuræ Sylvii* (S') communicates at 1 (Fig. II) with the *s. interparietalis* (ip), and in the manner spoken

OBSERVATION II.

Madarász, János—Confirmed Thief.

(Sclavonian.)

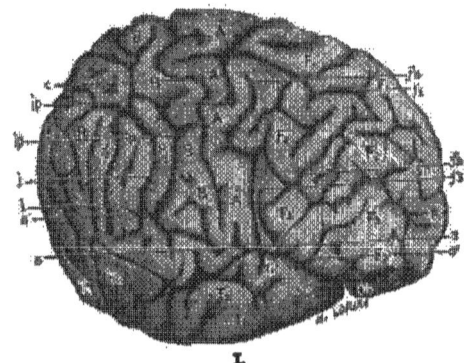

I.

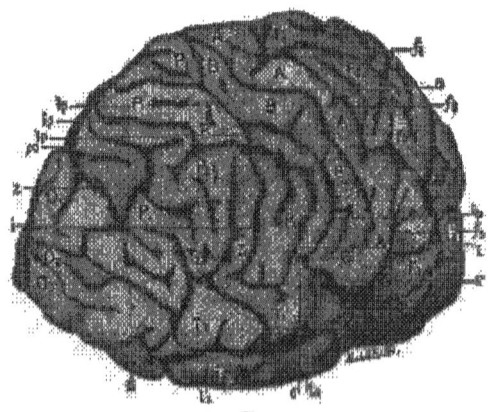

II.

OBSERVATION II.

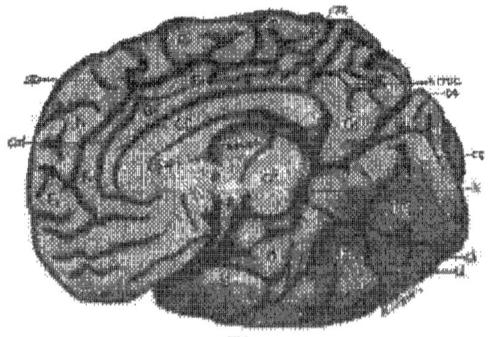

III.

A.—Gyrus Centralis Anterior.
B.—Gyrus Centralis Posterior.
Cu.—Cuneus.
CC.—Corpus Callosum.
F. 1.—Gyrus Frontalis Superior.
F. 2.—Gyrus Frontalis Medius.
F. 3.—Gyrus Frontalis Inferior.
Fs.—Gyrus Fusiformis.
Gf.—Gyrus Fornicatus.
H.—Gyrus Hippocampi.
Lg.—Gyrus Lingualis.
P. 1.—Lobulus Parietalis Superior.
P. 2.—Lobulus Parietalis Inferior.
P. 2'.—Gyrus Angularis (Lobulus Tuberis-Benedikt).
Q.—Lobulus Quadratus (Præcuneus).
S.—Fissura Sylvii.
S'.—Fissura Sylvii, Posterior Ramus.
S''.—Fissura Sylvii, Anterior Ramus.
T. 1.—Gyrus Temporalis Superior.
T. 2.—Gyrus Temporalis Medius.
T. 3.—Gyrus Temporalis Inferior.
U.—Gyrus Uncinatus.
O.—Occiput.
c.—Sulcus Centralis.
cc.—Fissura Calcarina.
cl.—Sulcus Collateralis.
cm.—Sulcus Calloso-Marginalis.
f 1.—Sulcus Frontalis Superior.
f. 2.—Sulcus Frontalis Inferior.
f. 3.—Sulcus Frontalis Perpendicularis.
h.—Scissura Hippocampi.
ho.—Fissura Horisontalis.
ip.—Sulcus Interparietalis.
po.—Fissura Parieto-Occipitalis.
s. cruc.—Sulcus Cruciatus.
t. 1.—Sulcus Temporalis Superior.
t. 2.—Sulcus Temporalis Inferior.
φ.—Secondary Sulcus Frontalis.

of, it can also be said that the posterior ascending branch of the *fissura Sylvii* communicates with the *f. parieto-occipitalis*.

The *s. temporalis superior* (t. 1) does not communicate with the *f. Sylvii*. (The connection seen in the photograph arose from a laceration of the specimen.) On the contrary, it unites with the *s. temporalis medius* (t. 2), and the united two send a branch (2, Fig. II) to the *s. occipitalis transversus* and the *f. parieto-occipitalis* (ho and po). (*This union, however, is not so extensive as it appears in the outline drawing, Fig. II of original work.*)

A blending of the *gyrus temporalis medius* (T. 2) with the *Lobulus tuberis* (P. 2') (*gyrus angularis* of authors) is prevented by a fissure.

The *f. parieto-occipitalis* (po) is not distinct from the *horizontalis* (ho), and it also connects with the *scissura hippocampi* (h, Fig. III). The *s. collateralis* (cl, Fig. III) scarcely reaches the *scissura hippocampi* (h).

The *gyri lingualis* and *fusiformis* (Lg and Fs) are directed almost entirely median-wise and slightly downward so that the cerebrum in the middle cranial fossa rests upon the *tentorium* more by an edge than by a surface.

Antero-posterior chord,	14.8
Hemispheric arch,	24.6
Anterior curve,	15.8
Middle curve,	3.7
Posterior curve,	5.1

LEFT HEMISPHERE.

The *s. centralis* (c) is in direct communication with the *f. Sylvii* (S) and anteriorly with the *s. frontalis superior* (f. 1) and is not entirely independent posteriorly of the *interparietalis* (ip). The *s. frontalis perpendicularis* (f. 3) upon the surface is not separated from the *f. Sylvii*. In the *gyrus frontalis superior* (F. 1) there are intercommunicating primary and secondary fissures and the vertical cross which communicates with the *s. centralis* may be looked upon as connected with the primary and secondary fissures. An interesting peculiarity of this cerebral hemisphere is, that the *f. Sylvii* (S) extends

OBSERVATION II. 43

to the point of the external border of the base of the frontal lobe and entirely defines the orbital lobe (Ob) from the gyri of the external and upper surface. The orbital lobe (Ob) has no *sulcus orbitalis cruciata*, but in its place a deep fissure which communicates also with the *f. Sylvii*. The *interparietalis* (ip) unites with both the *f. Sylvii* and the *s. occipitalis horizontalis* (ho).

The condition of the *s. temporalis superior* (t. 1) is peculiar. It ends in the *f. Sylvii* (S) just before this gives off the *ramus posterior* (S') and as the lower occipital fissure (g) rises from the first temporal fissure (t. 1) it is thereby indirectly united with the horizontal portion of the *f. Sylvii*.

The *f. parieto-occipitalis* (po) is separated from the *horizontalis* (ho) only by an extremely thin bit of gyrus. The *f. calcarina* (cc) is very short.

The connecting bit between the *gyri lingualis* and *fusiformis* (Lg and Gf) is bent at the medial border, to a right angle. In the same manner the transition of the medial basal lobes (H and U) is very abrupt, falling off through an anteriorly inclined wall, and as the occipital lobe is deeply bent, the basilar lobe (Lg and Fs) form a very sharply curved and downwardly directed surface.

The *gyri lingualis* and *fusiformis* appear altogether very short.

Antero-posterior chord,	14.6
Hemispheric arch,	26.4
Anterior curve,	13.6
Middle curve,	7.4
Posterior curve,	5.4

SKULL.[1]
Cubic contents, 138.0.

Horizontal circumf.,	.	51.0	Transv. and longit. circum.,		84.0
Ear circumf.,	.	31.2	Facial height,	.	11.3
Greatest length,	.	17.5	Frontal, . .	.	6.1
Greatest breadth,	.	14.7	Nasal, . . .	.	5.4
Frontal curve,	.	13.6	Ear and root-of-nose		
Parietal curve,	.	11.7	radius, .	.	11.4
Occipital curve,	.	10.3	Ear-occiput radius,	.	9.7
			Occipital shortening,	.	1.7

pfr—11.3 prfl—14.9 pfl—10.9 pllfr—14.0.[2]

Sagittal sutures and region of bregma partially obliterated, also the lower portion of the coronal suture, especially on the right. Also the sutures between the pterygoid processes of the sphenoid bone and the frontal bone are mostly effaced.

. A symmetric, rather markedly shortened, somewhat small and broad skull.

APPENDIX TO OBSERVATION II.

RIGHT: The *calloso-marginalis* (cm) communicates with the *s. cruciatus* of the *praecuneus* (Q) and this last with the *parieto-occipitalis* (po). (See Table II, Fig. III.)

The external orbital sulcus springs from the *fossa Sylvii* (S) and all three parts inter-communicate (at places very shallow). Orbital lobe (Ob) much dwarfed and for the most part covered by the middle basilar lobe (U and H).

LEFT: The *calloso-marginalis* (cm) is plainly separated from the fissure of the *praecuneus*, and these from the *parieto-occipitalis* (po), although not by broad intervals.

The orbital lobe (Ob) stunted: *fissura orbitalis* has not the cross-formation.

The external orbital fissure extends from the *fossa Sylvii* to the frontal extremity.

---

[1] I have taken a large number of chords and curves, only a few of which, more especially the characteristic ones, are mentioned.

[2] p. tuber parietalis. L tuber frontalis. r. right. l. left.

OBSERVATION III.

(TABLE III.)

KUSS (Johann), Servian; low grade of intelligence, taciturn, irascible individual, outrageous in language, and when at liberty, a drunkard. He shot and killed his son because he admonished him concerning his intemperance, and he also threatened to kill another son.

Cerebellum incompletely covered.

LEFT HEMISPHERE (TABLE III, FIGS. I–III).

S. centralis (c) full profile view, is not separated from the *f. Sylvii*. The *s. frontalis perpendicularis* (L 3) is in the same manner connected with the *f. Sylvii*.

The *gyrus frontalis superior* (F. 1) is divided through its middle part by the secondary fissure (φ). By this means there exists here a marked indication of the "*four convolution type.*"

The *s. frontalis perpendicularis* is composed of three parts, the lowest one of which represents the *sulc. front. perp.* (L 3), the central one the radiating branch of the *s. frontalis superior* (L 1), and the upper one the radiating branch of the secondary fissure (φ).

This last predominates over the *fissura frontalis superior*. The perpendicular frontal fissure is completely dissevered from the inferior frontal fissure (L 2), and the latter has at its anterior extremity a further radiating fissure. (See Fig. 1.)

The *sulcus interparietalis* (ip) is united with the *fissura Sylvii* (S and S′). There is also an extensive communication with the *sulcus temporalis superior* (t, 1) at 3.

OBSERVATION III.

Kuss, Johann—Murderer.
(Servian.)

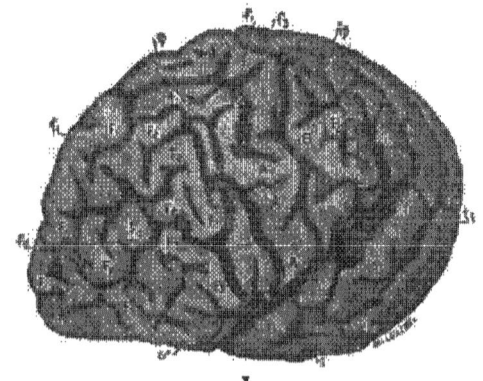

I.

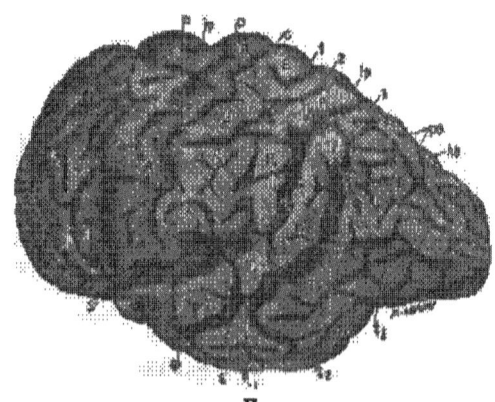

II.

OBSERVATION III.

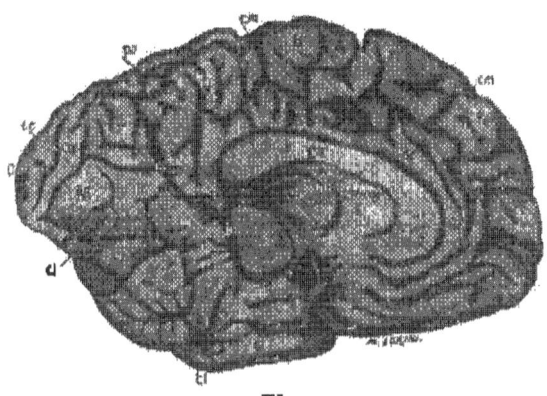

III.

A.—Gyrus Centralis Anterior.
B.—Gyrus Centralis Posterior.
Cu.—Cuneus.
CC.—Corpus Callosum.
F. 1.—Gyrus Frontalis Superior.
F. 2.—Gyrus Frontalis Medius.
F. 3.—Gyrus Frontalis Inferior.
Fu.—Gyrus Fusiformis.
Gf.—Gyrus Fornicatus.
H.—Gyrus Hippocampi.
Lg.—Gyrus Lingualis.
P. 1.—Lobulus Parietalis Superior.
P. 2.—Lobulus Parietalis Inferior.
P. 2'.—Gyrus Angularis (Lobulus Tuberis Benedikt).
Q.—Lobulus Quadratus (Præcuneus).
S.—Fissura Sylvii.
S'.—Fissura Sylvii, Posterior Ramus.
S''.—Fissura Sylvii, Anterior Ramus.
T. 1.—Gyrus Temporalis Superior.

T. 2.—Gyrus Temporalis Medius.
T. 3.—Gyrus Temporalis Inferior.
U.—Gyrus Uncinatus.
O.—Occiput.
c.—Sulcus Centralis.
cc.—Fissura Calcarina.
cl.—Sulcus Collateralis.
cm.—Sulcus Calloso-Marginalis.
f. 1.—Sulcus Frontalis Superior.
f. 2.—Sulcus Frontalis Inferior.
f. 3.—Sulcus Frontalis Perpendicularis.
h.—Scissura Hippocampi.
ho.—Fissura Horizontalis.
Ip.—Sulcus Interparietalis.
po.—Fissura Parieto-Occipitalis.
s. cruc.—Sulcus Cruciatus.
t. 1.—Sulcus Temporalis Superior.
t. 2.—Sulcus Temporalis Inferior.
φ.—Secondary Sulcus Frontalis.

The *sulcus temporalis superior* (t. 1) communicates at its rise (4, Fig. II) with the *Sylvii* (S), and at 3, as already mentioned, with the *s. interparietalis* (ip).

The *s. collateralis* (cl) communicates with the *scissura hippocampi* (h) and also with the *f. calcarina*. From the deep and many branched fissures in the *gyrus fusiformis* and *lingualis* (Fs and Lg) there result insular formations more numerous than is often offered for observation. The middle basilar lobe (H and U) is also extensively divided by transverse fissures from the *gyri fusiformis* and *lingualis*.

The *parieto-occipitalis* (po) is united with the *f. calcarina* (cc) and with the *scissura hippocampi*.

The *f. parieto-occipitalis* (po) is in direct communication with the *f. horisontalis* (ho), and so therefore with the *interparietalis*, superior temporalis (t. 1) and the sylvian (S').

Antero-posterior chord,	15.0
Hemispheric arch,	25.0
Anterior curve,	13.2
Middle curve,	5.2
Posterior curve,	6.6

RIGHT HEMISPHERE.

The *s. centralis* (c) is plainly separated from the *f. Sylvii* (S) by a thin bit of gyrus. In the upper third of both the anterior and posterior central gyri (A and B) are extensive depression formations.[1] ("Dellenbildungen.") The *gyrus frontalis superior* (F. 1) is divided by sagittal fissure formations into two gyri. The *s. interparietalis* is divided into an anterior and a posterior branch, the anterior branch not being

---

[1] In a triangular space marked off in front by the upper and middle frontal gyri and posteriorly by the upper half of the *gyrus centralis anterior*, there are two Y-shaped fissures. The upper one corresponds exactly to the usual depression of the upper third of the *gyrus centralis anterior* and by one of its branches connects with the central fissure (c). Both of the Y-fissures, with the stem and the anterior prong of the Y, represent in a greater or less degree, perpendicular fissures, which run parallel with the *sulcus frontalis perpendicularis*. One of these perpendicular branches extends barely to the *sulcus frontalis superior*, and the second (that is the upper one) that seems to result from the development of the depression ("delle") extends nearly to a secondary fissure of the *gyrus frontalis superior*.

distinctly separated from the *ramus posterior fissuræ Sylvii* and connects by two parallel branches with the *s. temporalis superior* (t. 1).

The anterior branch of the *interparietalis*, by means of an ascending branch, represents a *retrocentralis*, and here it is also clearly seen that it is partly accomplished through the development of fissures in the depression (delle) of the upper third of the *gyrus centralis posterior*. The *parieto-occipitalis* (po) is in direct communication on the one hand with the *occipitalis horizontalis* (ho), and on the other hand with the *scissura hippocampi* and *s. collateralis*.

The middle basilar lobe is but slightly fissured.

Antero-posterior chord,	16.4
Hemispheric arch,	23.6
Anterior curve,	12.6
Middle curve,	5.5
Posterior curve,	5.5

APPENDIX TO OBSERVATION III.

RUSS.

RIGHT: *s. cruciatus* of the *præcuneus* (Q) connects with the *parieto-occipitalis* (po), but not with the *calloso-marginalis* (cm).

It connects with a sagittal fissure which well separates the *præcuneus* from the *gyrus fornicatus*.

The *sulcus orbitalis* (ob) communicates with the *fossa Sylvii* (S).

The external orbital fissure is composed of two pieces: the anterior piece separates the base from the middle frontal gyrus (F. 2). The posterior portion is an incision from the *fossa Sylvii*. The middle part not distinct.

LEFT: The *calloso-marginalis* (cm) communicates through the fissures of the *præcuneus* (Q) with the *parieto-occipitalis* (po).

Moreover, the *calloso-marginalis* (cm) sends off a branch which well separates the *præcuneus* (Q) from the *gyrus fornicatus* (Gf). (See Table III, Fig. III.)

Of the external orbital fissure only the anterior part exists, which separates the base from the middle frontal gyrus (F. 2).

OBSERVATION IV.

(TABLE IV.)

PERUDINACE, Nicolaus, aet. 60, Sarvian peasant. Killed his son, who had advised him to live temperately.

The cerebellum is but just covered by the occipital lobes.

LEFT HEMISPHERE (TABLE IV, FIG. I-III).

The *s. centralis* (c) communicates at 6 with the *s. frontalis superior* (f. 1) as also with the inferior (f. 2) and perpendicular (f. 3). The *s. front. perpen.* (f. 3) is divided into two parts. The lower part, marked in the figure with f. 3, stands in the angle between the horizontal part (S) and the anterior ramus (S″) of the *fissura Sylvii*. The upper part constitutes the radiating branch of the *s. front. inferior* (L 2) and communicates with the *s. centralis* (c).

The lower portion of the *s. frontalis perpen.* (f. 3) unites with the *ramus anterior fissuræ Sylvii* (S″).

From the above mentioned two communications between the *s. centralis* and the frontal fissures, the *gyrus centralis anterior* (A) has lost its lobe character; it really consists of three separate pieces which represent looped extensions from each of the *gyri frontales* (Fig. I).

The *gyrus frontalis superior* (f. 1) has deep secondary fissures (g).

The upper third of the *gyrus centralis posterior* (B, Fig. II.) is poorly developed.

The *interparietalis* (ip) which is developed into a *retrocentralis*, is divided into anterior and posterior halves. At r,

OBSERVATION IV.

Praudinace, Nicolaus—Murderer.

(Servian.)

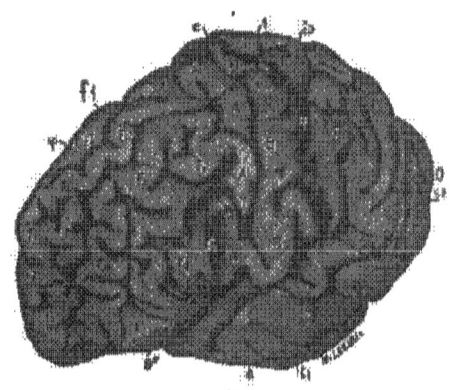

I.

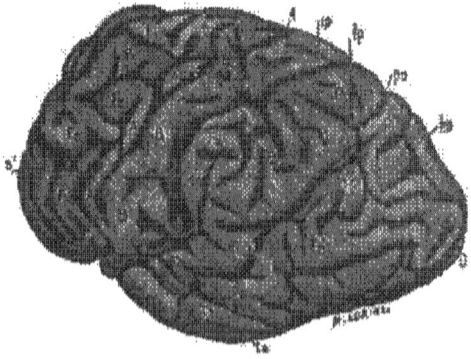

II.

OBSERVATION IV.

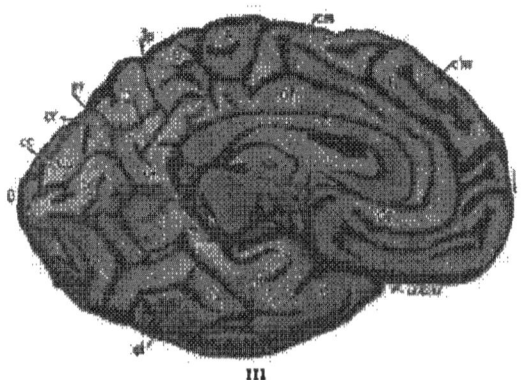

III

A.—Gyrus Centralis Anterior.
B.—Gyrus Centralis Posterior.
Cu.—Cuneus.
CC.—Corpus Callosum.
F. 1.—Gyrus Frontalis Superior.
F. 2.—Gyrus Frontalis Medius.
F. 3.—Gyrus Frontalis Inferior.
Fu.—Gyrus Fusiformis.
Gf.—Gyrus Fornicatus.
H.—Gyrus Hippocampi.
Lg.—Gyrus Lingualis.
P. 1.—Lobulus Parietalis Superior.
P. 2.—Lobulus Parietalis Inferior.
P. 2′.—Gyrus Angularis (Lobulus Toberis-Benedikt).
Q.—Lobulus Quadratus (Præcuneus).
S.—Fissura Sylvii.
S′.—Fissura Sylvii, Posterior Ramus.
S′′.—Fissura Sylvii, Anterior Ramus.
T. 1.—Gyrus Temporalis Superior.

T. 2.—Gyrus Temporalis Medius.
T. 3.—Gyrus Temporalis Inferior.
U.—Gyrus Uncinatus.
O.—Occiput.
c.—Sulcus Centralis.
cc.—Fissura Calcarina.
cl.—Sulcus Collateralis.
cm.—Sulcus Calloso-Marginalis.
f. 1.—Sulcus Frontalis Superior.
f. 2.—Sulcus Frontalis Inferior.
f. 3.—Sulcus Frontalis Perpendicularis.
h.—Scissura Hippocampi.
ho.—Fissura Horizontalis.
ip.—Sulcus Interparietalis.
po.—Fissura Parieto-Occipitalis.
s. cruc.—Sulcus Cruciatus.
t. 1.—Sulcus Temporalis Superior.
t. 2.—Sulcus Temporalis Inferior.
φ.—Secondary Sulcus Frontalis.

Fig. II, it sends a continuation over the medial border which at 1, Fig. III, enters the *s. calloso-marginalis* (cm), that is, into the *sulcus cruciatus* which unites with the *s. calloso-marginalis*.

The *s. temporalis superior* (t. 1) communicates by its initial portion with the *f. Sylvii* (S') and the posterior part of the *interparietalis* (ip) and indirectly also with the horizontal and perpendicular occipital fissures (ho and po).

The *perpendicularis* (po) communicates with the *horizontalis* (ho).

On the medial surface (Fig. III) the *s. calloso-marginalis* exists, composed of two non-communicating fissures (cm and c 1, m 1). The first, as before observed, connects with the *sulcus cruciatus* and at 1 with the *interparietalis* (ip).

The *parieto-occipitalis* (po) communicates with the *s. collateralis* (cl).

RIGHT HEMISPHERE.

The *s. centralis* (c) is separated from the *f. Sylvii* (S) only by a small bit of gyrus. The middle and upper thirds of the *gyrus centralis anterior* (A) consist of a simple connecting strip between the upper and middle *gyri frontales* (F. 1, F. 2). In the same manner also the lower third appears as a double twisted piece of gyrus from the *gyrus frontalis inferior* (F. 3). This lower third of the *gyrus centralis anterior* (A) instead of presenting a depression (delle) is divided by a deep fissure into anterior and posterior parts. Each one of these parts connects by a thin bit of gyrus with the lowest part of the *gyrus centralis posterior* (B). The upper connecting piece of it does not appear upon the surface, whilst the lower is represented by the piece of gyrus which separates the *s. centralis* (c) from the *f. Sylvii* (S).

The *s. centralis* (c) communicates with the *s. frontal perpen.* (f. 3) and also with the *s. frontalis inferior* (L 2). The *s. front. perp.* (f. 3) is not clearly separated on the surface from the *ramus anterior fissura Sylvii* (S"). A vertical branch from the upper frontal fissure forms the upper portion of a *præcentralis*. The *gyrus centralis posterior* (B) is slightly developed, especially in its upper two-thirds and has more the appearance of annectants than an independent gyrus. The *s. interparietalis* (ip) blends with the *f. Sylvii* (S) and does not extend to the *f. parieto-occipitalis* (po). The *interparietalis* forms a very pronounced *retrocentralis*.

The *s. temporalis superior* (t. 1) has only a very superficial communication with the *f. Sylvii* (S) and more with the *f. parieto-occipitalis* (po). By two branches running backwards from the upper temporal fissure (t. 1) the *gyrus temporalis medius* (T. 2) is prevented uniting with the *lobulus tuberis* (P. 2') (Gyrus Angularis). The *f. parieto-occipitalis* (po) is connected with the *f. horisontalis* only by a very shallow fissure. The *f. calcarina* (cc) very poorly developed.

The posterior branch of the *s. temporalis superior* (t. 1) blended with the *s. occipitalis inferior*, encircles at the base of the brain, the *lobuli lingualis* and *fusiformis* and connect with the *s. collateralis* and extends almost to the medial basal border. An extreme anterior part of this fissure here represents the *fissura fusiformis*. By an angular bent, deep fissure which unites with the *s. collateralis* (cl) the *gyri uncinatus* and *hippocampi* are rather sharply separated from the basilar occipital lobe. Upon this side the *s. calloso-marginalis* terminates more typically. It embraces the *gyrus paracentralis* fork-like, but with a lower branch it reaches the fissure between the *corpus callosum* and *gyrus corporis callosi* (*fornicatus*).

ON THE RIGHT: *gyrus orbitalis* short. An external orbital fissure, deep at places, separating the orbital gyrus from the frontal lobe communicates with the *f. Sylvii* and extends to the extreme anterior end of the frontal lobe. Parallel with this runs a short projection from the *f. Sylvii*.

ON THE LEFT: external orbital fissure composed of three parts and at places deep.

Antero-posterior chord,	14.9
Hemispheric arch,	23.2
Anterior curve,	13.0
Middle curve,	6.0
Posterior curve,	4.2

OBSERVATION IV.

SKULL.

Cubic contents, 87.2.

Horizont. circumf.,	52.5	Facial hight,	11.3
Ear—circumf.,	33.6	Frontal "	5.8
Greatest length,	18.0	Nasal "	5.5
Greatest breadth,	15.7	Ear—nasal-root radius,	11.9
Frontal curve,	13.0	Ear occipit "	9.1
Parietal "	13.5	Occipital shortening,	2.8
Occipital "	11.0		

The sutures throughout are almost effaced.

Extreme type of bracho-cephalic skull with great occipital shortening.

OBSERVATION V.

(TABLE V. FIG. I–III.)

SIRMA, Karl, aet. 48, Magyar, artist, bank-note counterfeiter.

Cerebellum barely covered by the occipital lobe.

RIGHT HEMISPHERE (FIG. I–III)

S. centralis (c) connects with *f. Sylvii* (S).

Middle third of *gyrus centralis anterior*, (A) through a deep incision of the superior frontal sulcus, (L 1) poorly developed.

Interparietalis (ip) connects with *f. occipitalis horizontalis* (ho). At 3, Fig. II, may be seen a deep operculose construction which necessarily impoverishes both upper thirds of the posterior central gyrus (B) and also the first and second parietal lobules (P. 1, P. 2). By a branch which extends backwards at 3, Fig. II, the *s. interparietalis* is put in shallow connection with the *ramus posterior fissuræ Sylvii* (S').

The upper temporal sulcus (t. 1) communicates at 2, Fig. I, with the *f. Sylvii* (S) and at 4, Fig. II, by a shallow fissure with the *f. occipitalis horizontalis* (ho) and the *interparietalis* (ip).

The *s. temporalis superior* (t. 1) by two branches (5 and 6, Fig. I) which extend backwards and unite in form of an arch, descends to the base and at 7, Fig. III, communicates with the *s. collateralis* (cl).

The *f. parieto-occipitalis* (po) is plainly separated from the *horizontalis* though only by a thin bit of gyrus.

The *f. parieto-occipitalis* (po) communicates by way of the

(57)

OBSERVATION V.

SINKA, KARL—Counterfeiter.

(Magyar.)

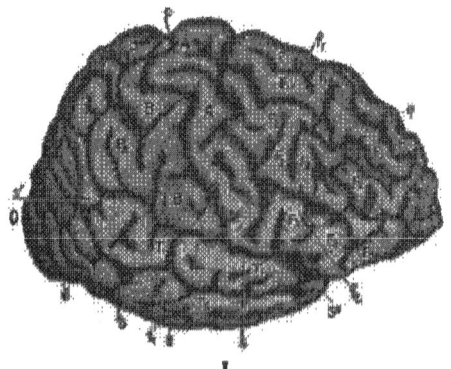

I.

II.

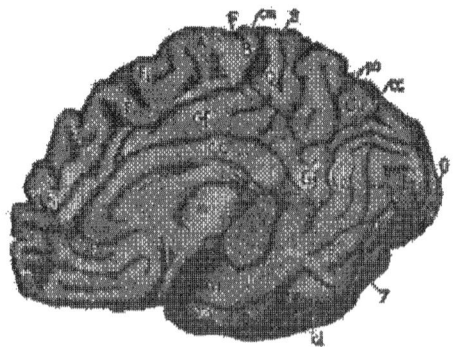

III.

A.—Gyrus Centralis Anterior.
B.—Gyrus Centralis Posterior.
Cu.—Cuneus.
CC.—Corpus Callosum.
F. 1.—Gyrus Frontalis Superior.
F. 2.—Gyrus Frontalis Medius.
F. 3.—Gyrus Frontalis Inferior.
Fs.—Gyrus Fusiformis.
Gf.—Gyrus Fornicatus.
H.—Gyrus Hippocampi.
Lg.—Gyrus Lingualis.
P. 1.—Lobulus Parietalis Superior.
P. 2.—Lobulus Parietalis Inferior.
P. 2′.—Gyrus Angularis (Lobulus Tuberis-Benedikt).
Q.—Lobulus Quadratus (Præcuneus).
S.—Fissura Sylvii.
S′.—Fissura Sylvii, Posterior Ramus.
S″.—Fissura Sylvii, Anterior Ramus.
T. 1.—Gyrus Temporalis Superior.
T. 2.—Gyrus Temporalis Medius.
T. 3.—Gyrus Temporalis Inferior.
U.—Gyrus Uncinatus.
O.—Occiput.
c.—Sulcus Centralis.
cc.—Fissura Calcarina.
cl.—Sulcus Collateralis.
cm.—Sulcus Calloso-Marginalis.
f. 1.—Sulcus Frontalis Superior.
f. 2.—Sulcus Frontalis Inferior.
f. 3.—Sulcus Frontalis Perpendicularis.
h.—Scissura Hippocampi.
ho.—Fissura Horisontalis.
ip.—Sulcus Interparietalis.
po.—Fissura Parieto-Occipitalis.
s. cruc.—Sulcus Crucialis.
t. 1.—Sulcus Temporalis Superior.
t. 2.—Sulcus Temporalis Inferior.
φ.—Secondary Sulcus Frontalis.

fissures of the *præcuneus* (Q) at 8, Fig. III, with the *s. calloso-marginalis* (cm). (This connection is plain in the photograph but is not represented in the outline of German Edition.) The anterior (k) and inferior (g) occipital fissures (Wernicke's f. and *s. occipito-temporalis*) are poorly developed.

The medial portion of the parietal lobe, that is, the *præcuneus* (Q, Fig. III) is impoverished by strong fissure formations, so that of the entire parietal lobe, the *lobulus tuberis (angularis)* (p. 2') is the only part not dwarfed.

The *cuneus* (cu, Fig. III) is similar in character to the *præcuneus* (Q).

The middle basilar lobe (U and H) is also impoverished to a high degree so that it seems only to furnish a point of junction between the *gyrus temporalis inferior* (T. 3) and the two gyri of the basilar occipital lobe (Fs and Lg). The separation of the middle and posterior basilar lobe is indicated.

Antero-posterior chord,	15.6
Horizontal arch,	21.5
Anterior curve,	12.5
Middle curve,	4.5
Posterior curve,	4.5

LEFT HEMISPHERE.

S. centralis (c) again ill-separated from the *f. Sylvii* and the two upper thirds of the *gyri centrales* barely developed.

The *s. frontalis perp.* (f. 3) blends with the *fossa Sylvii*. The *s. frontalis superior* (f. 1) exhibits at both its anterior and posterior ends decided operculose development, so that thereby the upper two-thirds of the *gyrus centralis anterior* (A) and also the *gyrus frontalis medius* (F. 3) are markedly dwarfed.

The *gyrus frontalis inferior* (F. 3) is greatly stunted. The *gyrus frontalis superior* (F. 1) in nearly its entire extent is divided into two parts so that we could really almost speak of four frontal gyri. Through a large vertical branch of the *s. frontalis superior* (f. 1) there results an upper part of a *præcentralis*.

The *s. interparietalis* (ip) shows a decided operculose formation from whence comes a poor development of the upper two-thirds of the *gyrus centralis posterior* (B), and also of the first and partly of the second parietal lobules (P. 1, and P. 2). The radial portion of the *interparietalis* does not join the sagittal.

The *s. temporalis superior* (t. 1) unites with the *interparietalis* (ip) through a deep fissure which enters the second parietal gyrus obliquely upwards and forwards and it is not clearly separated from the *f. occipitalis horisontalis* (ho).

Two curved branches of the *s. temp. sup.* (t. 1) running backwards and downwards are separated from that fissure by small bits of gyri, at the same time they separate the middle and posterior basilar lobe from the temporal and occipital lobes. The inferior occipital fissure (g) is united with Wernicke's *fissura fusiformis* (Fs).

The communication between the perpendicular and horizontal occipital fissures is in this case a very deep one. The *perpendicularis* (po) has no connection with the stunted *f. calcarina* (same as in apes). The middle basilar lobe (U and H) has a construction similar to the other side.

The *s. calloso-marginalis* (cm) of this side communicates also with the *f. parieto-occipitalis* (po) by way of the *s. cruciatus* (of the *præcuneus*).

External orbital fissure on both sides indistinct. Left orbital gyrus narrow.

Antero-posterior chord,	15.5
Hemispheric arch,	22.0
Anterior curve,	14.5
Middle curve,	3.05
Posterior curve,	4.04

OBSERVATION V.

SKULL—CONTENTS.

Horiz. circumf.,	52.0	Transv. and antero-post.	
Ear circum.,	31.0	circum.,	84.1
Greatest length,	17.6	Facial hight,	11.3
Greatest breadth,	14.8	Frontal hight,	5.6
Frontal curve,	13.4	Nasal hight,	5.7
Parietal curve,	10.0		
Occipital curve,	12.2		
H., (?)	12.8		

Ear nasal-spine radius r.	11.7:
l.	11.0:
Ear nasal-root radius r.,	11.9
l.	11.7
Ear-occipital radius,	10.6
Occipital shortening,	1.2
pfr—11.1; pfl—10.0; prfl—14.2; plfr—13.2.	

Sphenoidal sutures on both sides almost extinct. The sagittal suture also excepting its anterior portion; the coronal suture in places, and the lamboidal suture, especially on the left, in greater part obliterated.

Extreme type of bracho-cephalic skull with a high degree of asymetry of facial base and the tubera, and an extreme shortening of the parietal curve.

Corresponding with the shortening of the parietal curve the parietal lobes are extremely dwarfed.

OBSERVATION VI.

(TABLE VI.)

MAGLENOV, Gregory; aet. about 40, Servian, murdered a relation through revenge. Slightly developed intellect. As a prisoner, good natured.

The right cerebellum is not covered by the cerebrum and the outer border, on both sides is exposed (I have a photograph which exhibits this relation in the undivided encephalon).

RIGHT HEMISPHERE (TABLES VI, FIG. I-III).

The lower third of the *gyrus centralis posterior* (B) is completely divided by a fissure (at 4, Fig. II) and thereby a communication is established between the *s. centralis* (c) and the *f. Sylvii* (S).

Both central gyri are poorly developed. The *s. frontalis perpendic.* (f. 3) penetrates deeply into the base. The *s. frontalis superior* (f. 1) sends an extension into the *gyrus centralis anterior* (A) which extends almost to the *s. centralis* (c). The *s. frontalis inferior* (f. 2, see Fig. I) is composed of an anterior and a posterior part; the anterior connects with the perpendicular (f. 3) and the superior (f. 1) frontal sulci; the posterior with the *f. Sylvii* (S') and the latter unites with all three of the *sulci frontales* (Fig. 1).

The *interparietalis* (ip) connects with the *horizontalis* (ho), it emerges from the *fossa Sylvii* and at 2, Fig. II, communicates with the *s. temporalis superior* (t. 1).

The *s. temporalis superior* (t. 1) communicates not only with the *interparietalis* (ip) at 2, but also at 5, Fig. II, with the *f.*

OBSERVATION VI.

OBSERVATION VI.
MAGLENOV, GREGOR—Murderer.
(Servian.)

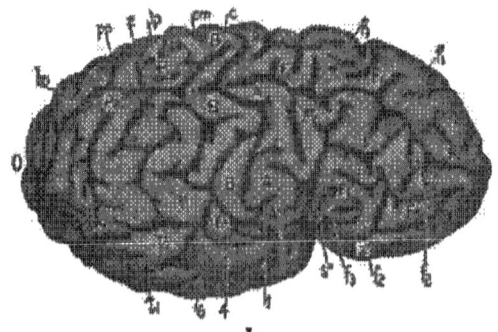

I.

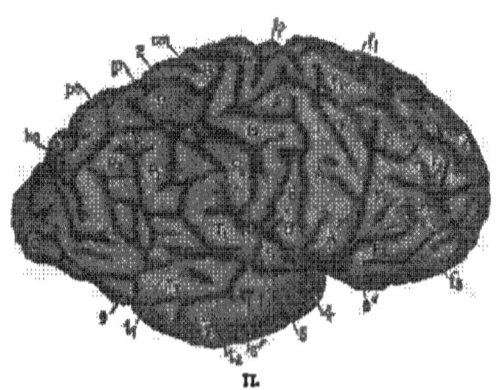

II.

OBSERVATION VI.

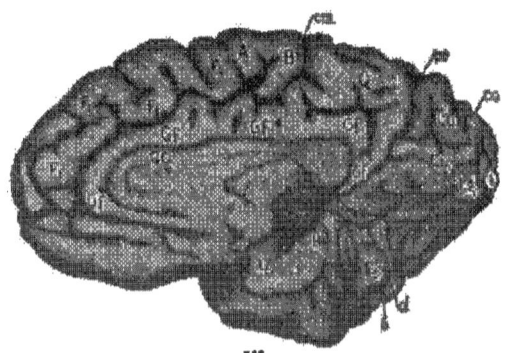

IIL

A.—Gyrus Centralis Anterior.
B.—Gyrus Centralis Posterior.
Cu.—Cuneus.
CC.—Corpus Callosum.
F. 1.—Gyrus Frontalis Superior.
F. 2.—Gyrus Frontalis Medius.
F. 3.—Gyrus Frontalis Inferior.
Fs.—Gyrus Fusiformis.
Gf.—Gyrus Fornicatus.
H.—Gyrus Hippocampi.
Lg.—Gyrus Lingualis.
P. 1.—Lobulus Parietalis Superior.
P. 2.—Lobulus Parietalis Inferior.
P. 2'.—Gyrus Angularis (Lobulus Tuberis-Benedikt).
Q.—Lobulus Quadratus (Præcuneus).
S.—Fissura Sylvii.
S'.—Fissura Sylvii, Posterior Ramus.
S''.—Fissura Sylvii, Anterior Ramus.
T. 1.—Gyrus Temporalis Superior.

T. 2.—Gyrus Temporalis Medius.
T. 3.—Gyrus Temporalis Inferior.
U.—Gyrus Uncinatus.
O.—Occiput.
c.—Sulcus Centralis.
cr.—Fissura Calcarina.
cl.—Sulcus Collateralis.
cm.—Sulcus Calloso-Marginalis.
f. 1.—Sulcus Frontalis Superior.
f. 2.—Sulcus Frontalis Inferior.
f. 3.—Sulcus Frontalis Perpendicularis.
h.—Schisura Hippocampi.
ho.—Fissura Horizontalis.
ip.—Sulcus Interparietalis.
po.—Fissura Parieto-Occipitalis.
s. cruc.—Sulcus Cruciatus.
t. 1.—Sulcus Temporalis Superior.
t. 2.—Sulcus Temporalis Inferior.
φ.—Secondary Sulcus Frontalis.

Sylvii. Besides this, it sends several branches backwards and downwards, one of which, (3, Fig. II) by a tortuous, shallow way, Fig. III, reaches the *s. collateralis* (c 1).

The *parieto-occipitalis* (po) is united to the *horizontalis* (ho. Fig. I and II), also by shallow fissures with the *scissura hippocampi* (h, Fig. III) and through deep fissures with the *s. cruciatus*.

The *s. collateralis* (c. 1) has also a shallow connection with the *scissura hippocampi* (h, Fig. III).

The middle basilar lobe is extremely dwarfed and is transversely separated from the posterior basilar lobe.

Antero-posterior chord,	15.9
Hemispheric arch,	21.0
Anterior curve,	11.7
Middle curve,	4.3
Posterior curve,	5.0

LEFT HEMISPHERE.

S. centralis (c) incompletely separated from the *f. Sylvii* (S) and unite with the *retrocentralis*.

The *s. frontalis perpend.* (L 3) arises from the *fossa Sylvii* and with the vertical branch of the *sulcus frontalis superior* (L 1) forms a *præcentralis*.

The *gyrus frontalis sup.* (F. 1) is divided in its entire length, so that there results a complete "four-convolution" type.

Frontal cerebrum stunted.

The *interparietalis* (ip) is divided into two pieces. The anterior piece, combined with the fissure system of the depression of the upper third of the *gyrus centralis posterior* (B), forms a *retrocentralis* which communicates with the *s. centralis* (c).

The *s. temporalis superior* (t. 1) indistinctly developed.

The *s. occipitalis inferior* (g) flows into the *fissura fusiformis* (Fs) and so together they separate the outer surface of the temporal and the occipital lobe from the middle and posterior basilar lobe. This common fissure makes a very shallow communication with the *s. collateralis* (cl) which is shallow.

The *f. parieto-occ.* (po) is in direct communication with the *horizontalis* (ho) and also with the *scissura hippocampi* and has a shallow connection with the *s. cruciatus* of the *præcuneus* (Q).

A separation of the middle from the posterior basilar lobe indicated.

The paracentral lobe well defined.

LEFT: There are three parts of an external orbital fissure, the anterior one of which separates the orbital gyrus (ob) from the *gyrus frontalis medius* (F. 2); the posterior and middle parts separate the *gyr. front. med.* from the *gyrus frontalis inferior* (F. 3). The posterior one constitutes a third lower branch of the *f. Sylvii*. RIGHT: External orbital fissure only indicated.

Antero-posterior chord,	15.7
Hemispheric arch,	21.0
Anterior curve,	11.2
Middle curve,	5.6
Posterior curve,	4.2

In the skull, upon dissection, I found a high degree of asymetry in the posterior fossa.

OBSERVATION VII.

PAUNOVICH, THEODOR—Murderer.

(Servian.)

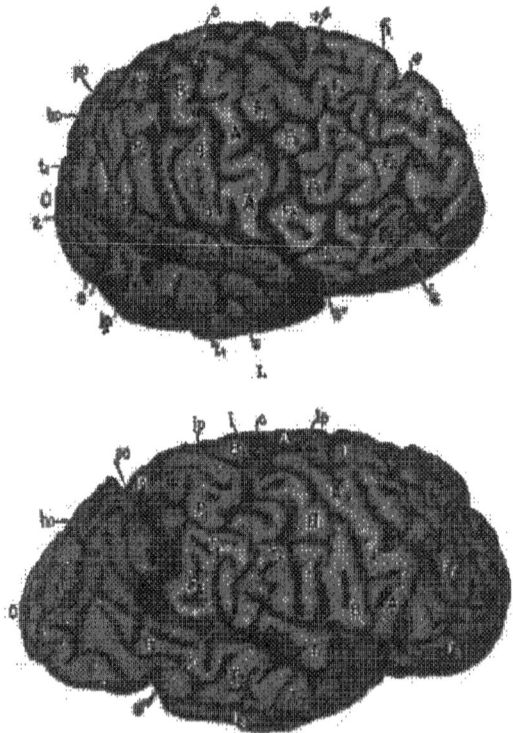

I.

II.

OBSERVATION VII.

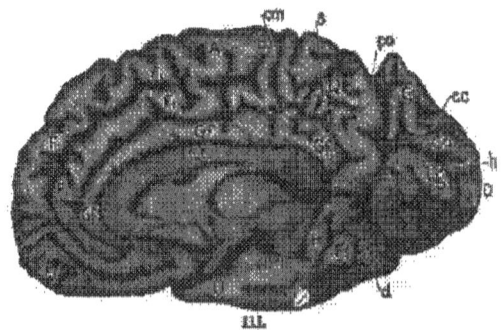

A.—Gyrus Centralis Anterior.
B.—Gyrus Centralis Posterior.
Cu.—Cuneus.
CC.—Corpus Callosum.
F. 1.—Gyrus Frontalis Superior.
F. 2.—Gyrus Frontalis Medius.
F. 3.—Gyrus Frontalis Inferior.
Fu.—Gyrus Fusiformis.
Gf.—Gyrus Fornicatus.
H.—Gyrus Hippocampi.
Lg.—Gyrus Lingualis.
P. 1.—Lobulus Parietalis Superior.
P. 2.—Lobulus Parietalis Inferior.
P. 3.—Gyrus Angularis (Lobulus Tuberis-Benedikt).
Q.—Lobulus Quadratus (Praecuneus).
S.—Fissura Sylvii.
S'.—Fissura Sylvii, Posterior Ramus.
S''.—Fissura Sylvii, Anterior Ramus.
T. 1.—Gyrus Temporalis Superior.

T. 2.—Gyrus Temporalis Medius.
T. 3.—Gyrus Temporalis Inferior.
U.—Gyrus Uncinatus.
O.—Occiput.
c.—Sulcus Centralis.
cc.—Fissura Calcarina.
cl.—Sulcus Collateralis.
cm.—Sulcus Calloso-Marginalis.
f. 1.—Sulcus Frontalis Superior.
f. 2.—Sulcus Frontalis Inferior.
f. 3.—Sulcus Frontalis Perpendicularis.
h.—Scissura Hippocampi.
ho.—Fissura Horizontalis.
ip.—Sulcus Interparietalis.
po.—Fissura Parieto-Occipitalis.
s. croc.—Sulcus Cruciatus.
t. 1.—Sulcus Temporalis Superior.
t. 2.—Sulcus Temporalis Inferior.
φ.—Secondary Sulcus Frontalis.

OBSERVATION VII.

(TABLE VII)

PAUKOVICKS, Theodor, Servian, aet. 30; formerly a laborer. After a protracted night-quarrel in an inn, he shot his antagonist with a gun which he procured for that purpose. Slight intellectual development.

Left cerebellum incompletely covered.

RIGHT HEMISPHERE (TABLE VII, FIG. I-III).

S. centralis (c) does not connect with *f. Sylvii* (S). The upper two-thirds of the *gyri centrales* (A and B) poorly developed. On account of the upper third of the *gyrus centralis posterior* (B) descending below the level of the gyrus, there occurs at i, Fig II, a junction of the *s. centralis* (c) with the *interparietalis* (ip).

The *s. front. perpen* (f. 3) communicates with the *f. Sylvii* (S) and is developed into a *præcentralis*.

The *gyrus frontalis medius* (F. 2) is entirely insulated and the *gyrus frontalis superior* (F. 1) is by means of two fissures (g. 4) divided into two gyri (Fig. I). The *interparietalis* (ip, Fig. II) which forms a well marked *retrocentralis* is not clearly separated from the *f. Sylvii* (S) (Fig. I), and as before mentioned, unites at i, Fig. II with the *s. centralis* (c) and is connected also with the *horisontalis* (ho, Fig. II).

The *s. temporalis superior* (t. 1) communicates at 2, Fig. I, with the *f. Sylvii* (S') and extends over into the broad *horisontalis* (ho, Fig. I).

Wernicke's fissure (k) and the lower occipital fissure (g)

communicate with each other and are well formed. The *horizontalis* (ho) which has an enormous depth, joins, deep down, with the *parieto-occipitalis* (po) (Fig. I).

Viewed from beneath, the middle basilar lobe does not rise much above the anterior one. The *gyri lingualis* and *fusiformis* imperfectly separated. The *f. parieto-occipitalis* (po) extends deeply into them, so that the entire occipital lobe hangs as it were on a thin inner stem. The *s. calloso-marginalis* (cm) connects at S. with the *scissura hippocampi* (h) by way of the *s. cruciatus* of the *præcuneus*. The border between the middle and posterior basilar lobe is sharp and ascends steeply at the rear.

Antero-posterior chord,	.	.	.	.	15.7
Hemispheric arch,	.	.	.	.	23.0
Anterior curve,		.	.	.	14.0
Middle curve,		.	.	.	3.0
Posterior curve,		.	.	.	6.0

LEFT HEMISPHERE.

S. centralis (c) does not communicate with *f. Sylvii* (S), but in place of this it communicates with the *interparietalis* (ip) and the *s. frontalis superior* (f. 1). The *s. frontal. perpend.* (f. 3) connects through the *s. frontalis inferior* with the *ramus anterior fissura Sylvii* (S') and forms a well defined *præcentralis*.

The *interparietalis* (ip) forms a *retrocentralis* which communicates, as before remarked, with the *s. centralis* (c), *f. Sylvii* (S), and *temporalis superior* (t. 1). The *f. parieto-occipitalis* (po) is poorly separated from the *horizontalis* (ho) and this last incompletely from the *s. temporalis superior* (t. 1).

The lower occipital fissure (g) and the *fissura fusiformis* (t. 3) are joined to each other but are not well developed. They have a double connection with the *s. collateralis* (cl). Here, as upon the other side, the separation of the middle from the posterior basilar lobe is indicated by a short transverse branch of the *s. collateralis* (cl). The *calloso-marginalis* (cm) communicates with the *f. parieto-occipitalis* (po).

The *fissura orbitalis* (ob) on both sides is very complicated,

An external orbital fissure as the boundary between the orbital gyrus (Ob) and the outer frontal lobe, exists on both sides. The right *fissura orbitalis* is composed of three parts, the posterior one emanating from the *fossa Sylvii*.

On the left the external orbital fissure is formed by three parts which have shallow communications with each other, and the posterior part communicates with the *fossa Sylvii*.

Antero-posterior chord,	16.0
Hemispheric arch,	24.3
Anterior curve,	13.5
Middle curve,	5.3
Pocterior curve,	5.5

OBSERVATION VIII.

(TABLE VIII).

FACTUSA, Gipsy, condemned thief.

Protruding occipital lobes of the cerebellum, especially upon the right side.

RIGHT HEMISPHERE (TABLE VIII, FIGS. I-III).

The *s. centralis* (c) (See photograph of Fig. I) is not separated from the *f. Sylvii* (S) and at 1, unites with the *s. frontal. perpend.* (f. 3) and, through a depression in the *gyrus centralis posterior* (B), with the *interparietalis* (ip). The *s. front. perp.* (f. 3) forms a *præcentralis*, as does the *interparietalis* a *retrocentralis*.

The middle and lower frontal gyri are separated from each other only in their anterior parts.

The *interparietalis* (ip) joins at 2, Fig. I, with the *f. Sylvii* and is in connection with the *f. occipitalis horisontalis* (ho).

The upper temporal sulcus (t. 1) has no communication with the sylvian fissure, as the fissures at 4 and 5 (Fig. II) are very shallow. The connection of the *interparietalis* (ip) with the *horisontalis* (ho) is in reality entirely superficial.

The middle temporal gyrus (T. 2) is separated from the *lobulus tuberis* (P. 2') (*Gyrus angularis*?).

The *f. parieto-occipitalis* (po) is not well separated from the *horisontalis* (ho) and *interparietalis* (ip) and is connected with the *scissura hippocampi* (h). This last is the case also (6, Fig. III) with the *s. collateralis* (cl).

OBSERVATION VIII.

FACZUNA, ZIGEUNER—Confirmed Thief.
(Gipsey.)

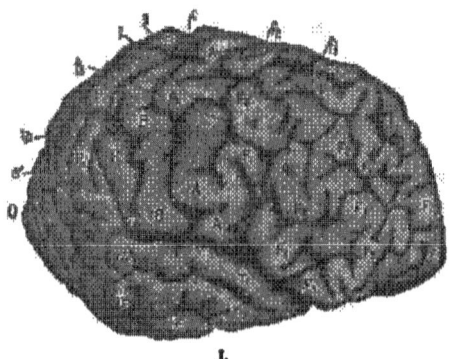

I.

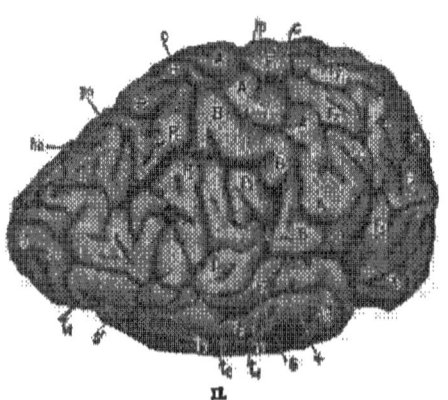

II.

OBSERVATION VIII.

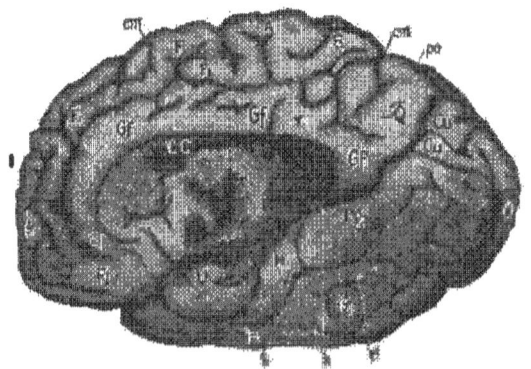

III.

A.—Gyrus Centralis Anterior.
B.—Gyrus Centralis Posterior.
Cu.—Cuneus.
CC.—Corpus Callosum.
F. 1.—Gyrus Frontalis Superior.
F. 2.—Gyrus Frontalis Medius.
F. 3.—Gyrus Frontalis Inferior.
Fu.—Gyrus Fusiformis.
Gf.—Gyrus Fornicatus.
H.—Gyrus Hippocampi.
Lg.—Gyrus Lingualis.
P. 1.—Lobulus Parietalis Superior.
P. 2.—Lobulus Parietalis Inferior.
P. 2'.—Gyrus Angularis (Lobulus Tuberis-Benedikt).
Q.—Lobulus Quadratus (Præcuneus).
S.—Fissura Sylvii.
S'.—Fissura Sylvii, Posterior Ramus.
S''.—Fissura Sylvii, Anterior Ramus.
T. 1.—Gyrus Temporalis Superior.

T. 2.—Gyrus Temporalis Medius.
T. 3.—Gyrus Temporalis Inferior.
U.—Gyrus Uncinatus.
O.—Occiput.
c.—Sulcus Centralis.
cc.—Fissura Calcarina.
d.—Sulcus Collateralis.
cm.—Sulcus Calloso-Marginalis.
f. 1.—Sulcus Frontalis Superior.
f. 2.—Sulcus Frontalis Inferior.
f. 3.—Sulcus Frontalis Perpendicularis.
h.—Scissura Hippocampi.
ho.—Fissura Horizontalis.
ip.—Sulcus Interparietalis.
po.—Fissura Parieto-Occipitalis.
s. cruc.—Sulcus Cruciatus.
t. 1.—Sulcus Temporalis Superior.
t. 2.—Sulcus Temporalis Inferior.
φ.—Secondary Sulcus Frontalis.

OBSERVATION VIII.

The occipital basilar lobe (Lg and Fs) is entirely flat and the apex of the occiput is on a level with the middle basilar lobe. The boundary between the middle and under basilar lobe is indicated by transverse fissures.

Wernicke's fissure (k) as well also as the *fusiformis* (fs) are well developed.

The *s. calloso-marginalis* (cm) thoroughly separates the *gyrus fornicatus* (Gf) from the *præcuneus* (Q).

Antero-posterior chord,	16.6
Hemispheric arch,	25.0
Anterior curve,	14.0
Middle curve,	5.0
Posterior curve,	6.0

LEFT HEMISPHERE.

The *s. centralis* (c) communicates with the *f. Sylvii* (S) and the upper frontal sulcus (f. 1). The perpendicular frontal sulcus (f. 3) arises from the *f. Sylvii*. A *sulcus frontalis inferior* (f. 2) is not distinct.

The *interparietalis* constitutes a well constructed *retrocentralis* and connects with the *horizontalis* (ho). It also communicates with the *f. Sylvii* (S) and is imperfectly separated from the *s. centralis* (c).

In its lower section, the upper temporal sulcus (t. 1) separates into two parts, the lower one of which communicates with the *f. Sylvii*.

The *f. parieto-occ.* (po) is well separated from the *horizontalis* (ho) but imperfectly from the *interparietalis* (ip). Wernicke's fissure (k) is united with the upper temporal sulcus (t. 1); is well developed and connects with a slight expression of an inferior occipital fissure, (g) *s. fusiformis* only indicated.

The occipital basilar lobe (Fs—Lg) occupy almost the same level as the middle lobe, and the position of this is exceptionally—(as seen from beneath)—deeper than the anterior portion.

The *s. calloso-marginalis* (cm) is well separated from the fissure of the *præcuneus* (Q).

OBSERVATION VIII.

RIGHT: External orbital fissure exists without communication with *f. Sylvii*.
LEFT: This fissure is composed of two parts, the posterior one of which arises from the *f. Sylvii*.

Antero-posterior chord,	16.9
Hemispheric arch,	24.0
Anterior curve,	14.3
Middle curve,	4.2
Posterior curve,	5.5

SKULL—CONTENTS, 1,500 CM.

Horizon. circum.,	52.0
Ear circumference,	32.0
Greatest length,	18.3
Greatest breadth,	14.0
Frontal curve,	12.4
Parietal curve,	12.8
Occipital curve,	12.0
Transverse and Antro-posterior Circum.,	76.5
Facial hight (Typical for Gipsy),	10.0
Frontal hight,	6.0
Nasal hight,	5.0
Ear nasal-root radius,	11.0
Ear occipital radius,	11.1
Occipital shortening (? F.),	0.1

No asymetry; slightly oxycephalic (pointed). Posterior portion of sagittal suture obliterated. Typical gipsy skull, somewhat oxycephalic.

OBSERVATION IX.

BUDIMCIC, LUKAS——Robber and Murderer.

(Servian.)

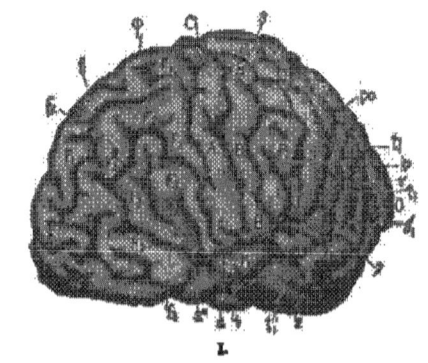

I.

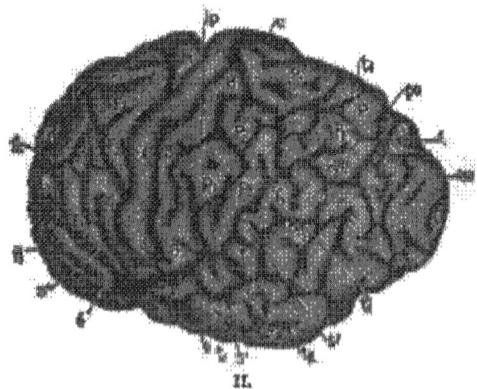

II.

OBSERVATION IX.

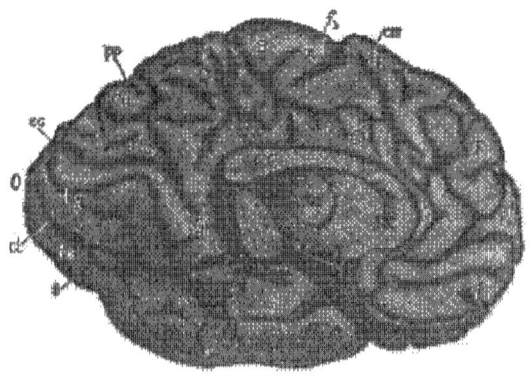

III.

A.—Gyrus Centralis Anterior.
B.—Gyrus Centralis Posterior.
Cu.—Cuneus.
CC.—Corpus Callosum.
F. 1.—Gyrus Frontalis Superior.
F. 2.—Gyrus Frontalis Medius.
F. 3.—Gyrus Frontalis Inferior.
Fu.—Gyrus Fusiformis.
Gf.—Gyrus Fornicatus.
H.—Gyrus Hippocampi.
Lg.—Gyrus Lingualis.
P. 1.—Lobulus Parietalis Superior.
P. 2.—Lobulus Parietalis Inferior.
P. g'.—Gyrus Angularis (Lobulus Tuberis-Benedikt).
Q.—Lobulus Quadratus (Præcuneus).
S.—Fissura Sylvii.
S'.—Fissura Sylvii, Posterior Ramus.
S''.—Fissura Sylvii, Anterior Ramus.
T. 1.—Gyrus Temporalis Superior.

T. 2.—Gyrus Temporalis Medius.
T. 3.—Gyrus Temporalis Inferior.
U.—Gyrus Uncinatus.
O.—Occiput.
c.—Sulcus Centralis.
cc.—Fissura Calcarina.
cl.—Sulcus Collateralis.
cm.—Sulcus Calloso-Marginalis.
f. 1.—Sulcus Frontalis Superior.
f. 2.—Sulcus Frontalis Inferior.
f. 3.—Sulcus Frontalis Perpendicularis.
h.—Scissura Hippocampi.
ho.—Fissura Horizontalis.
ip.—Sulcus Interparietalis.
po.—Fissura Parieto-Occipitalis.
s. cruc.—Sulcus Cruciatus.
t. 1.—Sulcus Temporalis Superior.
t. 2.—Sulcus Temporalis Inferior.
φ.—Secondary Sulcus Frontalis.

OBSERVATION IX.

(TABLE IX, FIG. I-III).

RUDINCIC, Lukas; aet. 37, Servian; highway robber and murderer, incapable of education.

Cerebellum uncovered by occipital lobes, especially on the right.

LEFT HEMISPHERE (TABLE IX, FIG. I-III).

The *s. centralis* (c) at 4, fig. I, communicates with *f. Sylvii*.
The *gyri centrales* (A. B.) composed of thin folds.

The *s. frontalis perpend.* (L 3) communicates indirectly at 6, Fig. II (through the *s. front. medius* (L 2) with the anterior branch of the *f. Sylvii* (S").

The third frontal fissure (L 3) is developed into a *praecentralis*. In figs. I and II, the third frontal sulcus (L 3) is seen to be in connection with an upper radical fissure which represents a vertical branch of a secondary fissure (φ) in the upper frontal gyrus (F) (compare Fig. I, Table IX with Fig. I, Table III).

The upper and middle frontal gyri poorly developed (F. 1, F. 2).

The *interparietalis* (ip) (at 2, Fig. I) communicates with the *f. Sylvii* (S) and again also by means of a loop (at S', Fig. I) and by shallow fissures with the *f. occip. horison*. (ho). It is developed into a *retrocentralis* and is connected with *s. centralis* (c).

The upper temporal sulcus (t. 1) communicates (5, Fig. II) with the sylvian fissure and also (in the neighborhood of 2,

Fig. I) by a shallow fissure with the loop of the *interparietalis* (S. 1', Fig. I) and extends to the united *horizontalis* and *perpendicularis* (ho and po).

Besides this, the first temporal sulcus (t. 1) sends a branch (7, Fig. I) backwards which (at 8, Fig. III) extends into the *gyrus fusiformis* (Fs) and is in superficial communication with the *s. collateralis* (cl).

A second branch of the first temporal sulcus (t. 1) is a separating fissure (t. 3) between the temporal and middle basilar lobes (*Wernicke's fissura fusiformis*). The existence of a second temporal sulcus is only indicated (t. 2).

The *parieto-occipitalis* (po), which, upon the medial surface, is very short, communicates (Fig. III) with the *horizontalis* (ho).

The *s. collateralis* (cl) gives off branches from both sides which separate the middle from the posterior basilar lobes almost in their entire extent.[1]

The occipital (O) and middle basilar lobes (H and U) much dwarfed. The *calloso-marginalis* (cm) has but an extremely shallow communication with the fissures of the *præcuneus*, the præcuneal fissures posteriorly are well separated. On both sides the *calloso-marginalis* consists really of two imperfectly joined parts, the posterior one of which makes the curve around the paracentral lobe.

Antero-posterior chord,	15.1
Hemispheric arch,	22.0
Anterior curve,	12.0
Middle curve,	5.5
Posterior curve,	4.5

RIGHT HEMISPHERE.

S. centralis (c) communicates with *f. Sylvii* (S); both *gyri centrales* (A. B.) very thin, excepting, perhaps, the lower third of the posterior gyrus (B). The third frontal sulcus (f. 3) is again developed into a *præcentralis*. The first frontal gyrus

---

[1] An extension of the *s. collateralis* divides the middle basilar lobe into its two gyri (H and U).

(F. 1), especially in the anterior two-thirds of its outer and upper extent, is greatly stunted.

The radial portion of the *interparietalis* (ip), as a *retrocentralis*, extends quite parallel with the *s. centralis* (c) to the medial border and is separate from the sagittal portion. The sagittal portion unites with the *horizontalis* (ho) and communicates with the posterior ramus of the *s. Sylvii* (S') and with the *s. temporalis superior* (t. 1).

The first temporal fissure (t. 1) communicates with the *interparietalis* (ip); the *horizontalis* (ho) and the sylvian fissure (S). The last communication (with the *f. Sylvii*) is so situated that the upper temporal sulcus (t. 1) extends to the base and there enters the sylvian fissure. In this way it separates the superior temporal gyrus (T. 1) from the *gyrus uncinatus* in its entire extent and constitutes as it were, a split-off continuation of the *s. fusiformis* (fs).

There is no blending of the second temporal gyrus (T. 2) with the *lobulus tuberis* (*gyrus angularis?*) and there exists a pronounced *parieto-temporal operculum*. The *lobulus tuberis* (*gyrus angularis?*) has a distinct transverse cleft. The *parieto-occipitalis* (po) is not distinctly separated from the *horizontalis* (ho) and is likewise connected with the *scissura hippocampi* (h).

The lower occipital fissure (g) is greatly developed and connects with a short but deep fissure which corresponds to *Wernicke's fissura fusiformis*; it also sends a connection to the *s. collateralis*, which, by a transverse branch, extends very near to the stem of the *parieto-occipitalis* (po). A branch of the *fusiformis* divides the *gyrus fusiformis* from the *uncinatus*, although the *gyri uncinatus* and *hippocampi* upon the other side, enter into the occipital basal lobe very nearly upon a plane; upon this side the transition is very abrupt.

The paracentral lobe is marked by very deep fissures.

The *calloso-marginalis* (cm) and the *s. cruciatus* of the *præcuneus*, are each of them isolated.

The external orbital fissure is represented on both sides by two fissures; on the left side they are very well developed.

The orbital fissure (ob) communicates on both sides with the *fossa Sylvii* of the base.

Antero-posterior chord,	14.7
Hemispheric arch,	20.6
Anterior curve,	11.6
Middle curve,	4.5
Posterior curve,	4.5

SKULL—CONTENTS, 1,196 !

Horizontal circumference,	46.8!
Ear circumference, (18¼ inches),	30.0
Greatest length,	15.6 !
Greatest breadth,	13.9 !
Greatest hight,	13.1
Frontal curve,	11.2
Parietal curve,	11.8
Occipital curve,	9.6
Facial hight,	10.7
Frontal hight,	6.0
Nasal hight,	4.8
Ear nose-base radius,	10.8
Ear occipital radius,	9.5
Occipital shortening,	1.3
Transverse and Antero-posterior Circum.,	89.1 !

Sphenoidal sutures on both sides nearly without trace; coronary sutures, especially the one on the right, greatly obliterated and the same, in places, with the sagittal sutures. Squamous sutures poorly marked.

Excepting the greatest hight (13.1) and the corresponding ear circumference, this skull exhibits dimensions belonging to boyhood. It is also in a high degree bracho-cephalic (short), has a moderate occipital shortening.

OBSERVATION X.

Rozsa, Andreas—Robber.

(Magyar.)

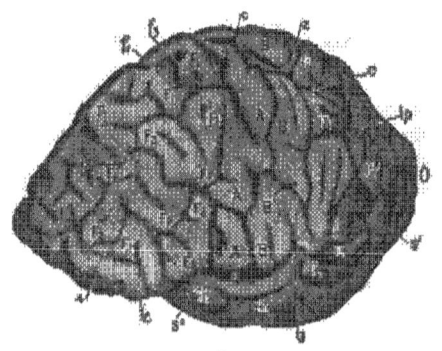

I.

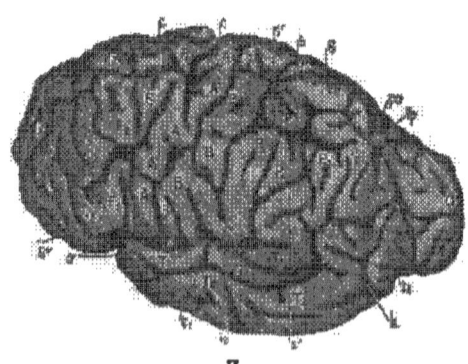

II.

OBSERVATION X.

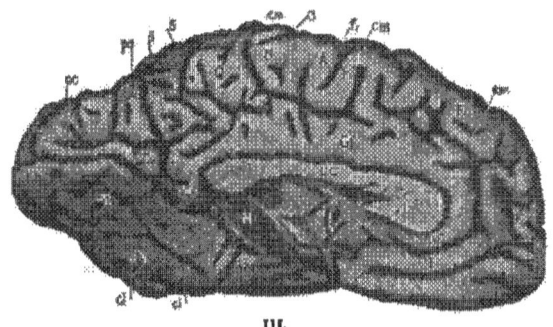

A.—Gyrus Centralis Anterior.
B.—Gyrus Centralis Posterior.
Cu.—Cuneus.
CC.—Corpus Callosum.
F. 1.—Gyrus Frontalis Superior.
F. 2.—Gyrus Frontalis Medius.
F. 3.—Gyrus Frontalis Inferior.
Fu.—Gyrus Fusiformis.
Gf.—Gyrus Fornicatus.
H.—Gyrus Hippocampi.
Lg.—Gyrus Lingualis.
P. 1.—Lobulus Parietalis Superior.
P. 2.—Lobulus Parietalis Inferior.
P. 2'.—Gyrus Angularis (Lobulus Taberis-Benedikt).
Q.—Lobulus Quadratus (Præcuneus).
S.—Fissura Sylvii.
S'.—Fissura Sylvii, Posterior Ramus.
S''.—Fissura Sylvii, Anterior Ramus.
T. 1.—Gyrus Temporalis Superior.
T. 2.—Gyrus Temporalis Medius.
T. 3.—Gyrus Temporalis Inferior.
U.—Gyrus Uncinatus.
O.—Occiput.
c.—Sulcus Centralis.
cc.—Fissura Calcarina.
cl.—Sulcus Collateralis.
cm.—Sulcus Calloso-Marginalis.
f. 1.—Sulcus Frontalis Superior.
f. 2.—Sulcus Frontalis Inferior.
f. 3.—Sulcus Frontalis Perpendicularis.
h.—Sulcus Hippocampi.
ho.—Fissura Horisontalis.
ip.—Sulcus Interparietalis.
po.—Fissura Parieto-Occipitalis.
s. croc.—Sulcus Cruciatus.
t. 1.—Sulcus Temporalis Superior.
t. 2.—Sulcus Temporalis Inferior.
φ.—Secondary Sulcus Frontalis.

OBSERVATION X.

(TABLE X.)

ROZSA, Andreas, aet. 68, Magyar, descended from a noted Family of robbers and was condemned for robbery. In prison he was good natured.

Cerebellum not covered by the occipital lobes.

LEFT HEMISPHERE (SEE TABLE X, FIG. I-III).

The *s. centralis* (c, Fig. I) unites with the *f. Sylvii* (S) and at 1 (Fig. I) with the third frontal sulcus (f. 3) as also at 2 and 3 with the *interparietalis* (ip).

The third frontal sulcus (f. 3) is connected by a shallow fissure with the ascending anterior ramus (S″) of the *f. Sylvii*.[1]

By a vertical branch (4, Fig. I) which connects the second frontal sulcus (f. 2) with the first (f.), the *gyrus frontalis medius* (F. 2) is cut into two parts, each part of which is almost an island by itself.

The upper frontal sulcus (f. 1) has a strong vertical branch which penetrates deeply into the *gyrus centralis anterior* (A) and represents an upper but separated piece of a *praecentralis*.

The greatly narrowed *gyrus frontalis superior* (F. 1) exhibits deep but short secondary fissures.

The *interparietalis* (ip) is developed into a well marked *retrocentralis* and is joined to the *horisontalis* (ho). The *retrocentralis* contains in its upper part an *operculum* by which

---

[1] In Fig. I (of outline sketch) the ascending ramus (S″) of the *f. Sylvii* is interrupted. The photograph shows the actual condition. (The illustration of the translation is correct).

the upper third of the *gyrus centralis posterior* (B) and the anterior portion of the upper parietal lobule (P. 1) are much reduced. The parietal lobule contains still another *operculum*.

The *interparietalis* (ip) communicates, as before observed, at 2 and 3, with the *s. centralis* (c) and in this wise indirectly with the *f. Sylvii* (S').

At 7 (Fig. II) the *s. temporalis superior* (t. 1) communicates with the *f. Sylvii* (S).

The temporal lobes, on both sides, seem stunted and at their expense the middle basilar lobes have a collossal development. The *gyrus hippocampi* (H) is extensively separated from the *gyrus lingualis* (Lg) and *gyrus uncinatus* (U) is clearly separated from the *gyrus fusiformis* (Fs), each by a transverse fissure.

The *s. temporalis medius* (t. 2) is only indicated, the third (t. 3) well developed and joined with Wernicke's fissure (k).

The *parieto-occipitalis* (po) is not in direct connection with the *horizontalis*, but by a very shallow fissure it communicates indirectly with it through the *interparietalis* (ip).

Upon the medial surface (Fig. III) the *parieto-occipitalis* (po) connects with the *s. cruciatus* of the *præcuneus* (Q). The *cruciatus* sends a branch which, upon the external surface (at 6, Fig. II), effects a shallow communication with the *interparietalis* (ip).[1]

The *parieto-occipitalis* (po) unites with the *scissura hippocampi* (h) and by a branch with the *s. collateralis* (c. 1).

The occipital basilar lobe (Lg and Fs) is very strongly and deeply fissured.

The *calloso-marginalis* (cm) connects (Fig. III) (in original outline sketch of German edition somewhat faulty) with fissure 6 and by this means communicates with the *parieto-occipitalis* (po) and *interparietalis* (ip). Besides this, it (cm) continues along the base of the *præcuneus* (Q) almost to the *parieto-occipitalis* (po) and thus separates in almost its entire extent, the *gyrus fornicatus* (Gf) from the *præcuneus* (Q).

---

[1] In outline sketch of original work, Fig. 6 is put too far forward ; (corrected in cuts of the Translation).

The orbital fissure (ob) communicates with the *f. Sylvii*, and upon both sides is of unusual depth.

Below the inferior one of the two anterior branches of the *f. Sylvii* (S), runs a fissure which separates both the inferior and middle frontal gyri (F. 3, F. 2) from the orbital lobe and communicates with the *s. frontalis inferior* (L 2).[1]

The occipital lobe dwarfed in its perpendicular dimensions. The paracentral lobule very completely defined.

Antero-posterior chord,	15.9
Hemispheric arch,	22.0
Anterior curve,	12.5
Middle curve,	3.0
Posterior curve,	6.5

RIGHT HEMISPHERE.

The *s. centralis* (c) is in communication with the *f. Sylvii* (S) not alone at its lower end but also by a fissure which breaks through the lower third of the *gyrus centralis posterior* (B) and ends in the sylvian fissure. The upper third of the posterior (B) and the middle third of the anterior (A) *gyri centrales* are composed only of thin pieces of gyri.

The third frontal sulcus (L 3) communicates with the *f. Sylvii*. It forms an extensive *præcentralis*, inasmuch as it not only connects with a vertical branch of the *s. frontalis superior* (L 1) but also with a vertical branch of a secondary fissure (v) of the *gyrus frontalis superior* (F. 1). The *gyrus frontalis medius* (F. 2) is divided into many islands through numerous little intercommunicating branches of the third, second, and first frontal fissures. The *gyrus frontalis superior* (F. 1) is much dwarfed.

The *interparietalis* (ip) is connected with the *horizontalis* and forms an extensive *retrocentralis*; it communicates with the *f. Sylvii* (S and S') in a two-fold manner and with the *s. temporalis superior* (t. 1) and it also has a shallow connection on the medial surface, with the *calloso-marginalis* (cm) by way of the *s. cruciatus* of the *præcuneus*.

The *s. temporalis superior* (t. 1) communicates with the *interparietalis* (ip). The gyrus which separates the *parieto-*

---
[1] This fissure corresponds to Broca's external orbital fissure in the Gorilla.

occipitalis (po) from the surrounding fissures on the upper surface, is strongly developed and a connection with the *interparietalis* (ip) is only indicated by a shallow fissure upon the just mentioned gyrus.

The disposition of the temporal lobes is peculiar.

The *gyrus temporalis superior* (T. 1) is divided into two parts by a deep fissure which communicates with the *f. Sylvii*.

The *gyrus temporalis medius* (T. 2) very narrow in the middle, broadens at the posterior end, and anteriorly coalesces with the *gyrus uncinatus* (U).

The second temporal sulcus (t. 2) separates the *gyrus temporalis medius* (T. 2) from the *gyrus fusiformis* (Fs).

The *gyrus fusiformis* is peculiarly formed, in that the lower occipital fissure (g) sends off a branch which not only divides the *gyrus fusiformis* into two parts, but penetrates deeply into the middle basilar lobe, separating it also into two gyri, which condition generally results only from a continuation of the *s. collateralis* (cl).

The *s. collateralis* (cl) joins with the *scissura hippocampi* (h). The occipital lobe in its perpendicular dimension extraordinarily dwarfed. The *s. orbitalis* in two-fold communication with the *f. Sylvii*.

An external orbital fissure is formed by a third anterior branch from the *f. Sylvii* and a small fissure between F. 2 and Ob.

Greatest length,	16.3
Greatest breadth,	7.3
Hemispheric arch,	21.0
Anterior curve,	12 5
Middle curve,	3.8
Posterior curve,	4.7

I saw three members of the Family and also the son and the nephew of the convict described in this observation. All three had markedly the same formation of head; the heads very high with flat back-heads, the surfaces of which were nearer to the transverse ear-line than were the frontal surfaces. I have in my possession a photograph of the subject of this observation.

OBSERVATION XL

Pantalic, Paul—Murderer.

(Servian.)

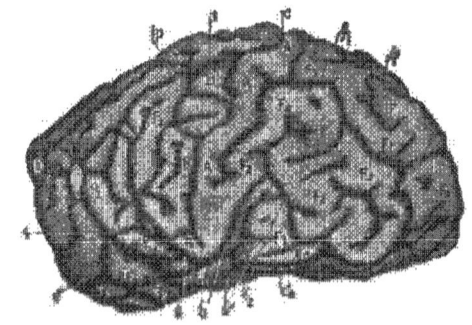

I.

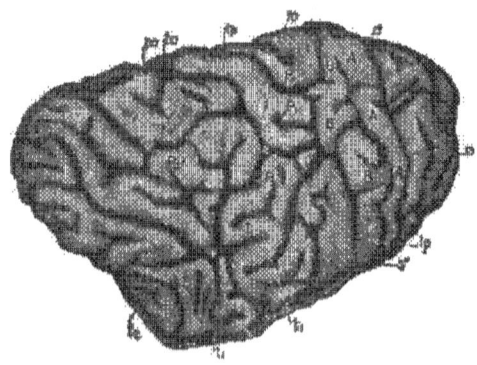

II.

OBSERVATION XL.

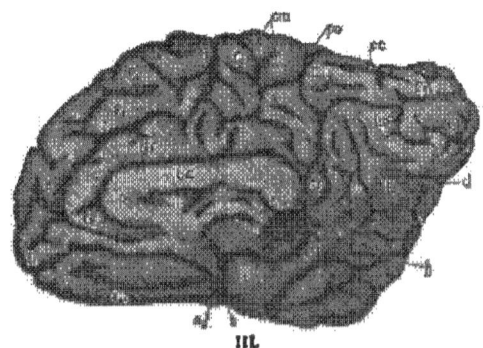

III.

A.—Gyrus Centralis Anterior.
B.—Gyrus Centralis Posterior.
Cu.—Cuneus.
CC.—Corpus Callosum.
F. 1.—Gyrus Frontalis Superior.
F. 2.—Gyrus Frontalis Medius.
F. 3.—Gyrus Frontalis Inferior.
Fu.—Gyrus Fusiformis.
Gf.—Gyrus Fornicatus.
H.—Gyrus Hippocampi.
Lg.—Gyrus Lingualis.
P. 1.—Lobulus Parietalis Superior.
P. 2.—Lobulus Parietalis Inferior.
P. 2'.—Gyrus Angularis (Lobulus Toberis-Benedikt).
Q.—Lobulus Quadratus (Præcuneus).
S.—Fissura Sylvii.
S'.—Fissura Sylvii, Posterior Ramus.
S''.—Fissura Sylvii, Anterior Ramus.
T. 1.—Gyrus Temporalis Superior.
T. 2.—Gyrus Temporalis Medius.
T. 3.—Gyrus Temporalis Inferior.
U.—Gyrus Uncinatus.
O.—Occiput.
c.—Sulcus Centralis.
cc.—Fissura Calcarina.
cl.—Sulcus Collateralis.
cm.—Sulcus Calloso-Marginalis.
f. 1.—Sulcus Frontalis Superior.
f. 2.—Sulcus Frontalis Inferior.
f. 3.—Sulcus Frontalis Perpendicularis.
h.—Schissura Hippocampi.
ho.—Fissura Horizontalis.
ip.—Sulcus Interparietalis.
po.—Fissura Parieto-Occipitalis.
s. cruc.—Sulcus Cruciatus.
t. 1.—Sulcus Temporalis Superior.
t. 2.—Sulcus Temporalis Inferior.
φ.—Secondary Sulcus Frontalis.

OBSERVATION XI.

(TABLE XI.)

PANTALIC, Paul, Servian; in company with the subject of observation XII, killed the husband of his priest's concubine, at the priest's instigation. The object was a small recompense.

The occipital lobes do not cover the cerebellum on either side, especially upon the left.

RIGHT HEMISPHERE (SEE TABLE XI, FIG. I–III).

The *s. centralis* (c. fig. 1) connects with *f. Sylvii* (S). The *central gyri* (A. B.) are very poorly developed. As the anterior and posterior rami (S″ and S′) of the *fissura Sylvii* meet at an acute angle, there really exists no horizontal portion of the *f. Sylvii*. The third frontal sulcus (f. 3) rises with the anterior ascending branch of the *f. Sylvii* from the angle made by this and the posterior branch (S).

Parallel with the third frontal sulcus (f. 3) runs a long vertical branch of the first frontal sulcus (f. 1) as the upper, but separate branch of a *præcentralis*.

The anterior part of the M. of the inferior frontal gyrus (F. 3) is extremely stunted.

The *gyrus frontalis medius* (F. 2) is very well developed; a deep fissure at g indicates the separation of the *gyrus frontalis superior* (F. 1).

The *interparietalis* (ip) connects with the perpendicular and horizontal occipital fissures (po and ho). It communicates besides, by fissures, shallow in places, (at 2, Fig. II) with the *s. temporalis superior* (t. 1) and at 3 (Fig. I) is only

OBSERVATION XL

slightly separated from the centralis (c). (*In the original German edition the outline sketch represents the connection too distinctly*).

At 4 (Fig. I) the *s. temporalis superior* (t. 1) communicates with the posterior ascending ramus of the *f. Sylvii* (S') and also at 2, as before stated, with the *interparietalis* (Ip) and at op (Fig. II) forms a complete *operculum*.

The *parieto-occipitalis* (po) connects directly with the *horizontalis* (ho) and thereby indirectly with the *s. temporalis superior* (t. 1, Fig. II).

On the medial surface (Fig. III) the *parieto-occipitalis* (po) unites with the *calloso-marginalis* (cm) by way of the *s. cruciatus* of the *præcuneus* (Q) and with the *scissura hippocampi* (h). In this wise the *gyrus fornicatus* is to a great extent separated from the *præcuneus*.

The *gyri lingualis* (Lg) and *fusiformis* (Fs) mount very abruptly from the basilar surface with no expressed *collateralis* (cl). The perpendicular diameter of the occiput extraordinarily diminished. The occipital basil lobe has no medial surface. A transverse fissure separates the *gyrus uncinatus* from the *gyrus fusiformis* (U and Fs).

The *s. occipitalis inferior* (g) and *Wernicke's fissura fusiformis* (t. 3) scarcely indicated, on the other hand Wernicke's fissure (k) is well developed.

The *gyrus ucinatus* (U) is extremely dwarfed and indeed represents only a blending of the three temporal gyri. Its entire sagittal length is 4.2 Cm. It does not rise above the anterior basilar lobe.

The orbital lobe (Ob) is developed to an unusually marked extent, and is separated from the frontal lobe of the external upper surface by a long fissure that comes as a third anterior branch from the *f. Sylvii*, indicated at S'' 2 (Figs. I and III) (similarity to animals). This fissure represents the under and middle part of the external orbital fissure; the anterior portion is indistinct.

Antero-posterior chord,	. . . .	15.0
Hemispheric arch,	. . . .	19.0
Anterior curve,	. . . .	9.5
Middle curve,	. . . .	4.5
Posterior curve,	. . . .	5.0

LEFT HEMISPHERE.

The *s. centralis* (c) communicates with the third frontal sulcus (f. 3) which is developed into a full *præcentralis*. The anterior part of the *gyrus frontalis inferior* (F. 3) poorly developed. The *s. frontalis superior* (L 1) communicates with the *s. f. medius* (L 2).

The *interparietalis* (ip) is not clearly separated from the *ramus posterior fissuræ Sylvii* (S'). A broad star-shaped fossa, with introverted annectant gyrus (*operculum parieto-temporal.*) prevents the union of the temporal gyrus with the *Lobulus tuberis* (*Gyrus angularis?*).

The *parieto-occipitalis* (po) on the upper surface is quite indistinctly separated from the *horisontalis* (ho) and the *interparietalis* (ip).

The border between the middle and posterior basilar lobes is indistinct and the posterior lobe is poorly developed. The two are completely separated by means of fissures.

The *calloso-marginalis* (cm) not distinctly connected with the fissures of the *præcuneus* (Q). At the base of the *præcuneus* there exists a fissure which separates it in almost its entire extent from the *gyrus fornicatus* (Gf).

The orbital lobe has a medium development, the posterior part without fissures.

Two little fissures which separate the third and second frontal gyri (F. 1, F. 2) from the base, but which do not communicate the *f. Sylvii*, represent the external orbital fissure.

OBSERVATION XL

SKULL—CONTENTS:

Horizontal circumference,	54.5
Ear circumference,	32.8
Greatest length,	18.7
Antero-posterior and horizont. circum.,	82.9
Greatest breadth,	15.5
Frontal curve,	14.2
Parietal curve,	12.8
Occipital curve,	12.8
Facial hight,	11.0
Frontal hight,	6.1
Nasal hight,	5.0
Ear and base of nose radius,	12.2
Ear occiput radius,	10.0
Occipital shortening,	1.2

There was found in the left lambdoidal suture a cunoid bone, 5 cm. long by 3.5 cm. broad.

A medium macrocephalic skull, moderate occipital shortening and moderate shortening of the parietal curve.

OBSERVATION XII.

OBSERVATION XII.
Mia, Michael—Murderer.
(Roumanian.)

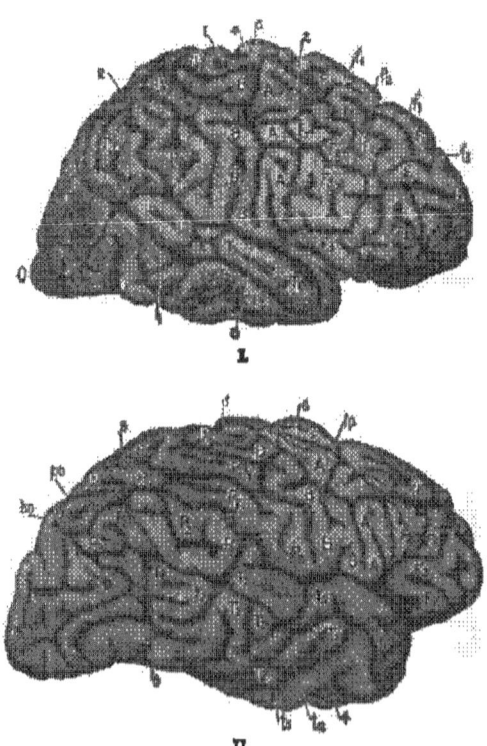

OBSERVATION XII.

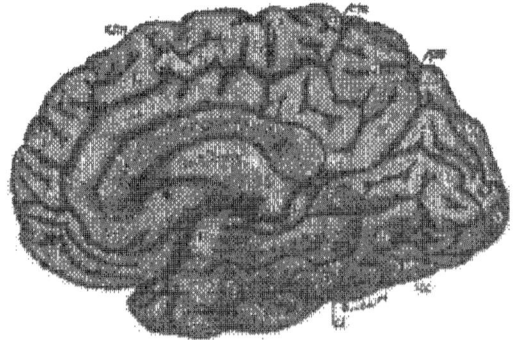

III.

A.—Gyrus Centralis Anterior.
B.—Gyrus Centralis Posterior.
Cn.—Cuneus.
CC.—Corpus Callosum.
F. 1.—Gyrus Frontalis Superior.
F. 2.—Gyrus Frontalis Medius.
F. 3.—Gyrus Frontalis Inferior.
Fs.—Gyrus Fusiformis.
Gf.—Gyrus Fornicatus.
H.—Gyrus Hippocampi.
Lg.—Gyrus Lingualis.
P. 1.—Lobulus Parietalis Superior.
P. 2.—Lobulus Parietalis Inferior.
P. 2'.—Gyrus Angularis (Lobulus Tuberis-Benedikti).
Q.—Lobulus Quadratus (Præcuneus).
S.—Fissura Sylvii.
S'.—Fissura Sylvii, Posterior Ramus.
S''.—Fissura Sylvii, Anterior Ramus.
T. 1.—Gyrus Temporalis Superior.

T. 2.—Gyrus Temporalis Medius.
T. 3.—Gyrus Temporalis Inferior.
U.—Gyrus Uncinatus.
O.—Occiput.
c.—Sulcus Centralis.
cc.—Fissura Calcarina.
cl.—Sulcus Collateralis.
cm.—Sulcus Calloso-Marginalis.
f. 1.—Sulcus Frontalis Superior.
f. 2.—Sulcus Frontalis Inferior.
f. 3.—Sulcus Frontalis Perpendicularis.
h.—Scissura Hippocampi.
bo.—Fissura Horizontalis.
Ip.—Sulcus Interparietalis.
po.—Fissura Parieto-Occipitalis.
s. cruc.—Sulcus Cruciatus.
t. 1.—Sulcus Temporalis Superior.
t. 2.—Sulcus Temporalis Inferior.
q.—Secondary Sulcus Frontalis.

OBSERVATION XII.

MIA, Michael; Roumanian; the companion of the subject of observation No. XI (Pantalio).

The massively-developed cerebellum is imbedded diagonally into the asymetrical basilar occipital lobe and in such a way that the right cerebellar hemisphere lays considerably more to the front than the left one did. For this reason the middle basilar cerebral lobe is compressed on the right and pushed forward.

LEFT HEMISPHERE.

The *s. centralis* (c) communicates with the first and third frontal sulci (f. 1 and f. 3).

The two lower thirds of both *gyri centrales* much reduced.

The *parieto-occipitalis* (po) is separated from the *horizontalis* (ho) by a very thin bit of gyrus.

The *lobulus tuberis* (*gyrus angularis?*) is separated into two parts by a long sagittal fissure. The continuation of the first and second temporal gyri (T. 1 and T. 2) into the second parietal gyrus (P. 2) and the *lobulus tuberis* (P. 2') is almost completely prevented by a group of fissures.

The anterior and middle part of an external orbital fissure well developed; posterior part not well developed.

RIGHT HEMISPHERE (SEE TABLE XII, FIGS. I–III).

S. centralis (c) not directly connected with the *f. Sylvii*. Is connected with the *interparietalis* (ip) at I (Figs. I and II).

The second frontal sulcus (f. 2) communicates by a branch (2, Fig. I) with the *s. centralis*.

By an extended fissure the upper frontal gyrus (F. 1) is divided into two parts.

The *interparietalis* (See ip, Fig. II) connects, as before observed, at I with the *s. centralis* (c) and with the *horizontalis* (ho) and is not separated from the posterior high-ascending branch of the *f. Sylvii* (S′). At 4 (Fig. II) the *superior temporal sulcus* (t. 1) communicates with the *f. Sylvii* and also at 5 (Fig. II) with the *interparietalis* (ip).

The *parieto-occipitalis* (po) does not connect with the *horizontalis* (ho). Instead of this, the *f. calcarina* (cc, Fig. III) connects with the *s. collateralis* (cl) and the *gyrus fusiformis* (Fs) is divided into a number of small islands.

The *calloso-marginalis* (cm) by a branch, separates the *gyrus fornicatus* (Fs) from the *praecuneus* (Q) and sends further in front a branch to the sulcus between the *gyrus fornicatus* and *corpus callosum* (CC).

The *parieto-occipitalis* (po) communicates with the fissures of the *praecuneus*.

On account of damage to the preparation nothing can be said concerning the external orbital fissure.

SKULL.—CONTENTS?

Horizontal circumference,	53.4
Ear circumference,	31.6
Greatest length,	18.1
Antero-posterior and horizont. circum.,	82.8
Greatest breadth,	15.0
Frontal curve,	13.2
Parietal curve,	12.0
Occipital curve,	11.2
Facial hight,	11.0
Frontal hight,	5.8
Nasal hight,	5.2
Ear—occiput radius,	9.5
Ear—base-of-nose radius,	12.3
Occipital shortening,	(?) 2.8!!
pfr.—parieto-frontal protuber. right,	10.8!
prfl.—parieto, right, fronto left protuberances,	14.9 (!)
pfl.—parieto frontal protub. left,	10.1 (!)
plfr.—parieto left, fronto-right protuberances,	14.0 (!)

Upper portion of frontal suture open. Lamboidal suture exhibits a rich development of small cuneiform bones.

Extreme occipital shortening, and asymetric skull.

OBSERVATION XIII.

BECZAR, Georg; aet. 35, Slovenian (Slovac) condemned for cruel lynch-law murder.

LEFT HEMISPHERE.

S. centralis (c) badly separated from the *f. Sylvii* (S) and *interparietalis* (ip) and communicates with the third frontal sulcus (f. 3).

The third frontal sulcus (f. 3) has a full communication with the *f. Sylvii* (S) and also connects, as just stated, with the *s. centralis*.

The condition of the frontal lobe (F) is especially interesting.

At first sight the upper frontal gyrus (F. 1) appears dwarfed and the middle one (F. 2) extraordinarily broad. Closer attention, however, shows the *præcentralis* to be composed of three radial branches. The lower one corresponds to the third frontal sulcus (f. 3) and the sagittal fissure belonging to it is the inferior frontal sulcus (f. 2). The middle portion of the *præcentralis* corresponds to the radial branch of the upper frontal sulcus (f. 1). It joins with the third frontal sulcus (f. 3).

The real upper frontal sulcus (f. 1) is shallow and stunted and composed of two sagittal parts, the posterior one of which is short and connects with the middle part of the *præcentralis*.

The sagittal fissure that at first glance appears as the upper frontal sulcus (f. 1) belongs to the upper portion of the (separated) *præcentralis*, therefore corresponds to the secondary fissure (v).

The middle frontal gyrus is thus not abnormally wide

nor the upper one unnaturally narrow, but the first one is extensively connected with the second, and the secondary gyrus has attained a great independence.

The *interparietalis* (ip) connects with the *f. Sylvii* (S) and penetrates so deeply into the posterior central gyrus (B) as to give it the appearance of being split.

The *s. temporalis* superior (t. 1) divides into two branches, one of which, by a shallow way, connects with the *horizontalis* (ho) and the *interparietalis* (ip) and the other divides the *lobulus tuberis* (P. 2′) (*gyrus angularis* ?) into two equal parts. A third branch first takes the form of an S as it descends, then rises upwards, then runs backwards and with its last part, as the *sulcus occipitalis inferior* (g) furnishes the under border of the *lobulus tuberis* (P. 2′) (*gyrus angularis* ?).

From the S-formed portion there extends a communication with the *s. collateralis* (cl).

Through the connection of the *fusiformis* (fs) with the *collateralis* (cl) the middle basilar lobe is entirely separated from the occipital basilar lobe, as the unusually deep and broad collateral sulcus communicates with the *scissura hippocampi* (h).

Wernicke's fissure (k) is a radial branch of the first temporalis (t. 1) and is well developed. The *s. occipitalis inferior* and *Wernicke's fissura fusiformis* (t. 3) are united and are in communication with the upper temporal sulcus (t. 1). The *fusiformis* separates the temporal gyrus from the *gyrus fusiformis*, but not from the *gyrus ucinatus*.

The superior temporal gyrus (T. 1) poorly developed. A separation of the underlying temporal surface into two gyri does not occur, and it is completely blended with the *gyrus ucinatus*.

The *parieto-occipitalis* (po) is well separated from the *horizontalis* (ho), but like the *s. collateralis* (cl) it communicates with the *scissura hippocampi* (h).

The *lobulus lingualis* (Lg) very stunted.

The fissures of the *præcuneus* (Q) are separated from other fissures, and the *præcuneus* is to a great extent separated from the *gyrus fornicatus*.

The posterior half of the stunted orbital lobe (Ob) runs

underneath the middle basilar lobe and the orbital fissure (Ob) connects with the *fossa Sylvii* (S). The external orbital fissure is composed first, of a third branch of the *f. Sylvii*, secondly, from a fissure which separates the M. of the lower frontal gyrus from the orbital gyrus. The anterior portion indistinct; the three parts separated from each other.

Antero-posterior chord,	16.8
Horizontal arch,	23.8
Anterior curve,	12.0
Middle curve,	7.0
Posterior curve,	4.8

RIGHT HEMISPHERE.

S. centralis (c) separated from the *f. Sylvii* (S) by a thin bit of gyrus.

The third frontal sulcus (f. 3) is not distinctly separate from the *f. Sylvii* (S), but is not connected with the *s. frontalis inferior* (f. 2).

Respecting the frontal lobe (F) there is the same condition as found upon the left side, except that its recognition is more difficult. The radial branch of the poorly-developed *s. frontalis superior* (f. 1) and that of the over-developed secondary fissure (φ) communicates with each other and with the third frontal sulcus (f. 3) thus forming a very marked *præcentralis*.

The *interparietalis* (ip) not connected with *f. Sylvii*; as a *retrocentralis* it penetrates deeply the upper part of the *gyrus centralis posterior* (B) to its depression (delle). It also is in communication with the *horisontalis* (ho) and independently with the *perpendicularis* (po). The upper temporal sulcus (t. 1) is in shallow communication with the *f. Sylvii* and its course is interrupted by a thin piece of gyrus. Its upper part communicates with the *interparietalis* (ip).

Wernicke's fissure (k) is unusually developed and forms a branch of the *s. temp. sup.* (t. 1).

At the place where these two fissures meet a fissure extends to the lower external border, which is the fissure that lies between the *gyrus fusiformis* and the temporal gyrus. Still higher up Wernicke's fissure sends off the *s. occipitalis inferior* (g).

The *parieto-occipitalis* (po) as before said, is latterly in connection with the *interparietalis* (ip). Notwithstanding this, the arched gyrus which surrounds the fissure is strongly developed.

The *parieto-occipitalis* (po) is also connected with the *scissura hippocampi* (h).

The *gyrus lingualis* (Lg) is greatly narrowed and is impaired, in its posterior part especially, by an unusually deep *operculose* formation, giving a deep chasm from which arises the *s. collateralis* (cl); this penetrates deeply into the middle basilar lobe, giving the basilar lobe the appearance of a loop from the *gyri lingualis* and *fusiformis*.

The gyrus orbitalis (Ob) is better developed than upon the other side; the orbital fissure (ob) the same as on other side. There is no third branch from the *f. Sylvii*. Middle part of external orbital fissure well developed; anterior part very shallow.

The *calloso-marginalis* (cm) is very peculiarly developed. In its anterior part it is composed of two parallel fissures and turns in front of the anterior central gyrus (A). The *central gyri* (A and B) are surrounded by a sharp curve, so that in this instance Betz' *lobus paracentralis* attains a rare degree of independence (upon the other side also this lobe is sharply defined).

From the fissure-arch which borders the *gyrus paracentralis* posteriorly, there extends a shallow fissure which well divides the *gyrus fornicatus* (Gf) from the *praecuneus* (π). The *s. cruciatus* is, as it were, split into fragments. On the right the occipital lobe is quite flat, on the left a little arched, so that the cerebellum has evidently an usually deep dip downwards.

Antero-posterior chord,	16.3
Hemispheric arch,	22.5
Anterior curve,	11.5
Middle curve,	4.5
Posterior curve,	6.5

OBSERVATION XIII.

SKULL—CONTENTS 1610 Ccm l

Horizontal circumference,	53.5
Ear circumference,	33.0
Greatest length,	18.2
Antero-posterior and transverse circum.,	84.6
Greatest breadth,	15.4
Greatest height,	12.8
Frontal curve,	13.3
Parietal curve,	13.4
Occipital curve,	11.5
Heighth of face (Facial height),	11.6
Heighth of forehead (Frontal height),	6.6
Heighth of nose (Nasal height),	5.2
Ear and base of nose radius,	11.0
Ear point of nose radius r.,	11.1
l.,	10.7 l
Ear occiput radius,	11.2
Occipital shortening,	0.2

Lamboidal sutures partially obliterated, especially at their confluence: the same case with the middle part of the squamous sutures of both sides.

Moderately macrocephalic skull with slight asymmetry of the facial base.

OBSERVATION XIV.

SCHENKER, Mathias, aet. 35, German vagabond and confirmed thief. Careless to an extreme degree, quick witted, ardent temperament.

The position of the cerebellum upon the arrival of the preparation, was peculiar. It was inclined forward. That this corresponded to a very peculiar position in life is evinced by the fact that the middle basilar lobe (U and H) on both sides were very nearly continuous with the occipital lobes (Lg and Fs), so that there was no cerebellar cavern to the hemispheres. On the skull the posterior part of the *foramum magnum* departs from the more normal horizontal position to a rather vertical one, and the posterior cerebral fossa is more vertical than horizontal.

LEFT HEMISPHERE.

The *s. centralis* (c) communicates with the *f. Sylvii* (S) and anteriorly there is a communication of which we shall speak further on.

The middle third of the *gyrus centralis anterior* (A) and the lower third of the *gyrus centralis posterior* (B) are well developed. The other parts are dwarfed.

The frontal lobe (F) is quite peculiar in its formation. The upper two-thirds of the *s. centralis* (c) is accompanied by a *praecentralis* which is made up from a combination of the vertical branches of the upper frontal sulcus (f. 1) and the secondary fissures (φ). These radial fissures are in no way connected with the sagittal. The upper vertical branch, however, sends a sagittal twig backwards, which, on the surface anterior to the *s. centralis* is not separated.

OBSERVATION XIV.

Back of this praecentralis runs another one, less parallel, and composed of the third frontal sulcus (f. 3) and of a second radial branch of the superior frontal sulcus (f. 1). Between the two *praecentrales* there exists, towards the medial border, a third radial fissure which is in connection with a long sagittal fissure. This last mentioned radial fissure together with its sagittal branch belongs to the system of secondary fissures (φ).

(There extends forward, moreover, from the second described *praecentralis* still another fissure, which, in its turn, sends upwards a radial branch that communicates with the secondary frontal fissure (φ)).

In order to understand the formation of these two marked praecentral fissures it must be remembered that when the upper frontal sulcus separates into two parts, each of these sagittal parts exhibits a tendency to form a vertical branch for itself. This is beautifully illustrated in the brain of the previous observation.

Here again the *gyrus frontalis superior* (F.1) is intergrown with the *gyrus fr. medius* (F.2), that is, they are not separated by any extensive sagittal fissure). The upper sagittal fissure is evidently a highly developed secondary fissure (φ).

The *s. frontalis inferior* (f. 2) is separated from its vertical branch, that is, from the third frontal sulcus (f. 3).

The *ramus anterior fissurae Sylvii* (S″) separates perpendicularly from the horizontal portion, and the horizontal part by a sharp downward curve extends much further forward than the point where the ascending branch leaves.

The *interparietalis* (ip), which is not, at the surface, distinctly separated from the *ramus posterior fissurae Sylvii* (S′) and which forms a pronounced *retrocentralis*, communicates with the *horisontalis* (ho) but this is not connected with the *parieto-occipitalis* (po).

The *lobulis tuberis* (*gyrus angularis?*) is not located on the upper external surface of the hemisphere but on the posterior surface. This, as will be easily comprehended, is still more the case with the occipital lobe, which does not appear on an external and upper surface, but rests upon a posterior face.

Wernicke's fissure (k) and the lower occipital sulcus are not distinctly developed; the temporal sulcus or *fissura fusiformis*, however, is.

The *parieto-occipitalis* (po) lies quite obliquely backwards and upwards and has a shallow communication with the *scissura hippocampi* (h). The *calloso marginalis* (cm) continues on to the vicinity of the *parieto-occipitalis* and separates to a great extent the *gyrus fornicatus* from the *praecuneus* (Q) and is in communication with the praecuneal fissures.

As before said, the brain at this point is quite abnormal in form, as the occipital part of the base (Fs and Lg) viewed from below, really rises higher than the middle part. The occipital basilar lobe is not visible from the medial view. The *collateralis* (cl) extends deeply into the middle basilar lobe. The *gyrus lingualis* exhibits very little fissuring, especially at its posterior part. The *gyrus fusiformis* very narrow. The *gyri uncinatus* and *hippocampi* short, narrow, and dwarfed by deep fissures. The stunted orbital lobe (Ob) faces more externally than inferiorly. The external orbital fissure is represented by two fissures, the posterior one of which is not distinctly separated from the *fossa Sylvii* (S).

Antero-posterior chord,	14.5
Hemispheric arch,	23.3
Anterior curve,	14.3
Middle curve,	5.5
Posterior curve,	3.5

RIGHT HEMISPHERE.

The *s. centralis* (c) connects with the *f. Sylvii* (S) and at two places with the *interparietalis* (ip) and at two points is but imperfectly separated from the perpendicular frontal sulcus (f. 3).

The perpendic. front. sulcus (f. 3) communicates with the *fossa Sylvii* (S), and as before remarked, with the *s. centralis* (c) and, as upon the other side of the brain, there exists between the real *praecentralis* and the *s. centralis* a deep parallel fissure which commences high up on the external surface and which evidently corresponds partly to a vertical branch of the superior frontal sulcus (f. 1) and partly to the radial branch of the secondary fissure (φ).

The second real *praecentralis* is a combination of the third frontal sulcus, (f. 3) and a vertical branch arising from the territory of the superior frontal sulcus (f. 1) and the secondary fissure (φ).

A third *praecentralis* is composed of the *ramus anterior fossae Sylvii* (S') and a vertical branch which belongs to the system of the superior frontal sulcus (f. 1). In the posterior part of the frontal lobe the secondary fissure (φ) in great measure takes the place of the superior frontal sulcus. The orbital lobe (Ob) is less dwarfed than on the left side. The external orbital fissure is the same as upon the other side, only more fully connected with the *fossa Sylvii*.

The entire parietal lobe lies almost at the posterior, very little on the upper, surface. The *interparietalis* (ip) is divided into two parts, of which the posterior communicates with the *horizontalis* (ho).

The temporal lobe, with the horizontal part of the *f. Sylvii*, extends far backwards towards the cerebral extremity. The superior temporal sulcus (t. 1) communicates by a very shallow fissure with the *f. Sylvii* (S).

The *gyrus temporalis superior* (T. 1) extraordinarily reduced. A second (t. 2) temporal sulcus not distinct. The *f. fusiformis* (t. 3) strongly developed.

The *parieto-occipitalis* (po) communicates with the *horizontalis* (ho) by means of the *interparietalis* (ip).

The *parieto-occipitalis* (po) also runs more diagonally. The *f. calcarina* (cl) is very short, so that the Cuneus is reduced to a minimum. The paracentral lobe is sharply defined by a fissure which does not communicate with the *calloso-marginalis* (cm).

The occipital portion of the base lies on a line with the middle part. The greater portion of the *gyrus lingualis* lies on the medial surface, inasmuch as the the *s. collateralis* (cl) runs near to and almost parallel with the medial border.

Antero-posterior chord, ... 15.0
Hemispheric arch, ... 22.6
   Anterior curve, ... 16.0
   Middle curve, ... 4. 2
   Posterior curve, ... 2. 4

SKULL OXYCEPHALIC (pointed).

Horizontal circumference, ... 50.0
Greatest length, ... 17.1
Greatest breadth, ... 14.0
Antero-posterior and transverse circum., . 81.8

Large pneumatic spaces in the frontal bone; great thickness of the occipital bone at its section through the *prom. max. occipitalis*. The facial aspect offers an unusual expression for European skulls from the fact that the height of the *fossae orbitales* equals their breadth.

OBSERVATION XV.

PROKETE, Peter, aet. 40; Hungarian, carpenter; unable either to read or write. A tall, powerful form with bristling blonde hair, prominent features, high cheek-bones, small piercing eyes. Up to his marriage was a robber and noted for his cruelty; since then, a professional thief. He was violent even in prison. Another prominent trait in him was his great love for his children.

This characteristic was mainly based, probably, upon a strong and active sexual passion which was fully satiated in his marital relations.

The cerebellum was moderate in size and not covered by the cerebrum.
LEFT HEMISPHERE.
S. centralis (c) separated from the *f. Sylvii* (S) by small bit of gyrus.

The third frontal sulcus (f. 3) very imperfectly separated from the *f. Sylvii* (S) and by a strong vertical branch united with the superior frontal sulcus (f. 1) so as to form a *praecentralis* which reaches over the entire external surface. The *gyri centrales* (A and B) very poorly developed.

Of the three frontal gyri, the middle one is the broadest and it is cut into anterior and posterior halves. The *interparietalis* (ip) communicates with the *f. Sylvii* (S) and extends uninterruptedly to the *horizontalis* (ho).

A branch of the *interparietalis* (ip) continues as a *retrocentralis* entirely to the medial surface at which locality it is not distinctly separated from the *calloso-marginalis* (cm). Also the *s. cruciatus* of the *praecuneus* (Q) communicates with the *interparietalis* (ip).

The upper parietal lobe (P. 1) poorly developed.

(110)

OBSERVATION XV.

The upper temporal sulcus (t. 1) has but very shallow communication with the *f. Sylvii* (S) though a deep one with Wernicke's fissure (k). There exists no well marked second temporal sulcus (t. 2) on the external surface. The second temporal gyrus (T. 2), which is thereby rendered very large, is separated from both the *gyrus fusiformis* (F 3) and *uncinatus* (U) by a fissure which extends almost from the occipital apex to the horizontal portion of the *f. Sylvii* (S) and in this manner makes a complete separation of the posterior and middle basilar lobes from the external surface, as is the case, for example, in the horse.

This *fissura basilaris lateralis* sends a transverse extension to the base which transversely separates the posterior from the middle basilar lobe and unites with *s. collateralis* (cl).

For professional readers, it need hardly be specified that this lateral basilar fissure consists of a blending of the inferior occipital sulcus with the fusiform fissure of Wernicke.

The *parieto-occipitalis* (po) is separated from the *horizontalis* (ho) by a thin piece of gyrus.

There is a connection (very shallow) between the *f. calcarina* (cc) and *s. collateralis* (cl). Both extend almost to the *scissura hippocampi* (h). The orbital lobe (Ob) is very broad; the orbital fissure (ob) communicates with the sylvian. The external orbital fissure is represented by an anterior and a middle portion. The *calloso-marginalis* separates the *praecuneus* (Q) from the *gyrus fornicatus*, communicates with the *s. cruciatus* of the *praecuneus* and extends nearly to the *parieto-occipitalis* (po).

Antero-posterior chord,	16.6
Hemispheric arch,	26.0
Anterior curve,	14.5
Middle curve,	5.3
Posterior curve,	6.2

RIGHT HEMISPHERE.

By a thin piece of gyrus the *s. centralis* (c) is distinctly separated from the *f. Sylvii* (S) but poorly so from the *interparietalis* (ip) at its foot. Both *gyri centrales* (A and B) very dwarfed.

The third frontal sulcus (L 3) connects with the *fossa Sylvii* (S). Notwithstanding the strong development of the vertical branch of the upper frontal sulcus (L 1) it does not connect with the third frontal sulcus (L 3) so that the *praecentralis* is composed of two pieces.

The anterior lobe (F) is small. The upper frontal gyrus (F. 1) is separated into two gyri by two very deep fissures so that the "four convolution type" is well expressed.

It can be seen that the anterior of these two secondary fissures has a radial branch which is connected with the vertical branch from the *s. frontalis superior* (L 1).

The *interparietalis* (ip) has a shallow connection with the foot of the *f. Sylvii* (S) and also a connection still further along with the *ramus posterior fissurae Sylvii* (S'); on the surface is not indistinctly separated from the upper temporal sulcus (t. 1). Furthermore, it connects with the *horizontalis* (ho).

The upper temporal sulcus (t. 1) connects with the *f. Sylvii* and, as before said, superficially with the *interparietalis* (ip) and the *occipito-horizontalis* (ho).

The temporal lobe is poorly developed, and a third temporal sulcus (t. 3) separates to a great extent the outer surface of the *gyrus fusiformis* (Fs) from the *gyrus uncinatus* (U). The *s. collateralis* (cl) sends upwards a Y-formed fork which is imperfectly separated from the *s. fusiformis*, and which separate in their entire extent the middle from the posterior basilar lobe.

The *s. collateralis* (cl), by means of a very shallow fissure, reaches the *f. calcarina* (cc).

Wernicke's fissure (k) is in connection with the *occipitalis horizontalis* (ho) and this with the *parieto-occipitalis* (po) so that there exists a complete "*ape-fissure*."

The *parieto-occipitalis* (po) as well as the *s. collateralis* (cl) connects with the *scissura hippocampi* (h).

The middle basilar lobe is extremely dwarfed.

The *paracentral* lobe is extraordinarily developed, and the *calloso-marginalis* (cm) is poorly separated from the upper frontal sulcus and extends through the fissures of the *prae-*

OBSERVATION XV.

cuneus far back into the vicinity of the *parieto-occipitalis* (po). External orbital sulcus is composed of the three typical parts. The *fissura orbitalis* (ob) communicates with the *f. Sylvii*.

Antero-posterior chord,	16.3
Hemispheric arch,	25.8
Anterior curve,	14.6
Middle curve,	6.0
Posterior curve,	5.2

SKULL—CONTENTS, 1,445 Ctm.

Horizon. circum.,	52.8
Ear circumference,	30.9
Greatest length,	18.5
Antero-posterior and horizont. circum.,	75.1 l
Greatest breadth,	13.9 l
Frontal curve,	13.6 l
Parietal curve,	11.2 l
Occipital curve,	11.8
Facial height,	12.4
Frontal height,	7.2
Nasal height,	5.5
Ear and spine of nose radius, r.	12.0 l
l.	11.7 l
Ear and root of nose radius, r.	12.0 l
l.	11.5 l
Ear occipital radius,	10.6
Occipital shortening,	1.1

Almost complete synosteosis of the sagittal suture. Lamboidal suture retains only islands like, coronary suture partly obliterated ; mastoideo-occipital suture obliterated.

Scaphocephalic, highly doliocephalic skull with notable shortening of the parietal curve, moderate occipital shortening and asymetry of the frontal and parietal bases.

OBSERVATION XVI.

PETRICEEWICZ, Peter, aet. 53; Croatian; tempted by money and the promise of marriage, he first sought to poison the husband of his paramour and then, in company with another, killed him. In prison he was quiet and penitent.

The cerebrum overlies the cerebellum, on the right about 2.5 Cm.: on the left about 1.7 Cm.

LEFT HEMISPHERE.

The *s. centralis* (c) connects with the *f. Sylvii* (S) and the third frontal sulcus (L 3). The *s. centralis* connects in a peculiar manner with a secondary fissure (φ) of the upper frontal gyrus (F. 1). The secondary fissure then extends to the medial surface and is in shallow communication with the *calloso-marginalis* (cm).

The upper frontal sulcus (f. 1) is composed of two parts, each of which has an important radial fissure. The two radial fissures lie parallel, one behind the other, and form the middle portion of the *praecentralis*, which has no communication with any other radial fissure.

The superior frontal gyrus (F. 1) contains four consecutive secondary fissures, each of which has a radial branch. The posterior one, as before remarked, connects with the *s. centralis* (c) and the *calloso-marginalis* (cm).

Its radial branch forms the upper part of the *praecentralis* and communicates with no other parts.

The secondary fissure lying next in front of the one just spoken of, has a radial branch which connects with the anterior radial branch of the *s. frontalis superior* (f. 1).

The *s. frontalis perpend.* (f. 3) as already mentioned, is connected with the *f. Sylvii* (S) and *s. centralis* (c).

The *gyri centrales* (A and B), especially the upper two-thirds of the anterior one and the upper third of the posterior, are poorly developed.

The *interparietalis* (ip), with the half of an upper vertical branch, constitutes a complete *retrocentralis*. It has a shallow connection with the horizontal *f. Sylvii* (S) and with the *ramus posterior fissurae Sylvii* (S'); on the other hand it has a deep connection with the upper temporal sulcus (t. 1). It extends to the *horizontalis* (ho) which in its turn reaches by a deep fissure to both the *perpendicularis* (po) and Wernicke's fissure (k).

The *s. temporalis superior* (t. 1) is in shallow communication with the *f. Sylvii*, in deep connection with the *interparietalis* (ip) and in direct connection with the *horicontalis* (ho).

The lower occipital sulcus (g) likewise originates from the upper temporal sulcus (t. 1); which last communicates several times with the *s. temporalis medius* (t. 2). The separating fissure between the temporal and middle basilar lobes (U and H) very slightly developed.

The *s. temporalis superior* (t. 1) possesses a deep operculose formation, from which connecting branches extend to all previously mentioned junctions.

The *gyri uncinatus* (U) and *hippocampi* (H) are strongly marked with fissures and a fissure separates the middle basilar lobe from the posterior.

Both the *f. parieto-occipitalis* (po) and *s. collateralis* (cl) extend to the *scissura hippocampi* (h). The *calloso-marginalis* (cm) has no posterior connections.

The fissure which separates the greatly reduced *praecuneus* (Q) from the *gyrus fornicatus* is connected with the *parieto-occipitalis* (po). The *s. cruciatus* of the *praecuneus* is poorly developed.

The *gyrus fornicatus* (Gf) has well marked transverse fissures.

The external orbital fissure, which separates the lower frontal gyrus (F. 3) from the base is well developed and communicates with the *f. Sylvii* (S).

An anterior separated portion which usually serves to divide the base from the *gyrus frontalis medius* (F. 2) is only indicated.

The entire brain, with a short base, seems relatively strongly arched.

Antero-posterior chord,	15.3
Hemispheric arch,	24.8
Anterior curve,	14.2
Middle curve,	4.8
Posterior curve,	5.8

RIGHT HEMISPHERE.

S. centralis (c) connects with *f. Sylvii* (S); *gyri centralis* (A and B) greatly reduced.

The third frontal sulcus (f. 3) is joined to a strongly developed vertical branch of the superior frontal sulcus (f. 1) from which results a well marked *praecentralis*. The third frontalis sulcus (f. 3) connects with the *f. Sylvii* (S). The *ramus anterior fissurae Sylvii* (S″) connects with all three of the frontal sulci.

The upper frontal gyrus (F. 1) is separated by deep secondary fissures into two parts. The *interparietalis* (ip) is broken into an anterior and a posterior part. It represents a complete *retrocentralis* and by its anterior portion is in communication (very shallow) with the horizontal part of the *f. Sylvii* (S) and with the *ramus posterior fissura Sylvii* (S′), also extensively with the *s. temporalis superior* (t. 1). By its very short posterior portion it communicates with the *horizontalis* (ho) and *perpendicularis* (po).

As already remarked, the *parieto-occipitalis* (po) communicates with the *horizontalis* (ho), Wernicke's fissure (k) and the *scissura hippocampi* (h).

The *s. temporalis superior* (t. 1) has only a slight connection with the *horizontalis* (ho) and gives off a lower occipital sulcus (g).

A third temporal sulcus makes a separation in the region of the temporal lobe (T), between the outer and under faces of the brain.

OBSERVATION XVI.

The *s. collateralis* (cl) has many branches and penetrates deeply into the poorly developed middle basilar lobe (U & H).

The *calloso-marginalis* is composed of an anterior and a posterior half. The latter surrounds the *lobus paracentralis*; has sundry connections with the fissure between the *gyrus fornicatus* and *corpus callosum* and is not separated from the *s. cruciatus* of the *praecuneus*. The *s. cruciatus* communicates with the *scissura hippocampi* (h) and is very indistinctly separated from a branch which the *parieto-occipitalis* (po) sends to the *praecuneus*.

An external orbital fissure, composed of two parts, separates not only the third (F. 3) but also the second frontal gyri (F. 2) from the orbital lobe (Ob). Its posterior part arises from the *fossa Sylvii*.

This hemisphere (right) is also strongly arched upon a short base.

Antero-posterior chord,	15.6
Hemispheric arch,	25.0
Anterior curve,	15.2
Middle curve,	5.3
Posterior curve,	4.5

SKULL—CONTENTS, 1,410 Cm.

Horizontal circumference,	51.3
Ear circumference,	31.4
Greatest length,	18.1
Antero-posterior and horizont. circum.,	78.4
Greatest breadth,	14.2
Frontal curve,	12.5
Parietal curve,	11.8
Occipital curve,	11.0
Height of face (Facial height),	11.1
Height of forehead (Frontal height),	6.3
Height of nose (Nasal height),	5.5
Ear and root of nose radius,	11.7
Ear—occiput radius,	10.5
Occipital shortening,	1.2

Sutures normal. Markedly dolichocephalic skull with medium occipital shortening.

OBSERVATION XVII.

(TABLE XVII.)

LOKSIK, Georg, Slovak; confirmed thief.

Both sides of cerebellum sufficiently covered.

LEFT HEMISPHERE.

The *s. centralis* (c) at its lower end approaches almost the horizontal portion of the *f. Sylvii* (S); it has only a shallow connection with it in its posterior lower third. In its middle third it is indistinctly separated from the *interparietalis* (ip).

Central gyri (A and B) poorly developed.

Third frontal sulcus (f. 3) is in superficial connection with the *ramus anterior fissura Sylvii* (S''). It connects with the radial branch of the upper frontal sulcus (f. 1) and forms a well defined *praecentralis*. At the point of junction the third frontal sulcus (f. 3) is ill-separated from the *s. centralis* (c).

The *interparietalis* (ip) forms a well developed *retrocentralis*, and is in three-fold communication with the *ramus posterior fissurae Sylvii* (S'). One of the three communications is made by means of the *horisontalis* (ho) up to which the *ramus posterior fis. Syl.* (S') reaches.

The *s. temporalis sup.* (t. 1) is divided into an under and an upper half, and is connected with the horizontal part of the *fissura Sylvii* (S) as well as with its posterior ramus (S'). Wernicke's fissure (k) is distinct and also a continuation of the *horizontalis*. The occipital lobe on the outer and upper surface reduced to a minimum.

The *gyri fusiformis* and *uncinatus* are separated by a long fissure (*fissura fusiformis*, t. 3) from the external surface. It is well developed and also takes the place of the lower occipital sulcus (g).

This *s. fusiformis* (t. 3) communicates with the *s. collateralis* (cl) and gives off a branch which effects a distinct separation of the *gyrus hippocampi* (H) from the *lingualis* (Lg).

The *parieto-occipitalis* (po) does not communicate with the *horizontalis* (ho) but instead of this it does with the *scissura hippocampi* (h).

The *interparietalis* (ip) extends deeply into the medial surface and is imperfectly separated from the *s. cruciatus* of the *praecuneus*. This sulcus is also in connection with the *parieto-occipitalis* (po) and communicates several times with the fissure between the *gyrus fornicatus* and *corpus callosum*. On the contrary, it is well isolated from the *calloso-marginalis* (cm), which last-named fissure is double.

The *s. orbitalis* (ob) communicates with the *f. Sylvii*. The external orbital fissure is indicated at several points.

Antero-posterior chord,	16.3
Hemispheric arch,	25.0
Anterior curve,	13.7
Middle curve,	5.5
Posterior curve,	5.8

RIGHT HEMISPHERE.

The *s. centralis* (c) is S-shaped and is ill-separated both from a vertical branch of the upper frontal sulcus (f. 1) and the third frontal sulcus (f. 3); indeed, the first connection is made by an enormous thinning and fold-like depression at the first point on the anterior central gyrus. A very shallow, superficial connection exists between the *s. centralis* (c) and *f. Sylvii* (S).

The third frontal sulcus (f. 3) is moreover in communication with the *ramus anterior fissurae Sylvii* (S').

Several fissures ascend from the *fossa Sylvii* (S), penetrating deeply into the frontal lobe (F), really representing a connection with all three of the frontal sulci.

A vertical branch of the upper frontal sulcus (f. 1) forms an upper but separate piece of a *praecentralis*, and this vertical piece connects with the *s. centralis* (c).

The *interparietalis* (ip) forms a *retrocentralis* along the upper third of the unusually ill-developed *gyrus centralis posterior* (B). The *interparietalis* is divided into an anterior and a posterior part, communicates somewhat indirectly with the *ramus posterior fissurae Sylvii* (S') and with the upper temporal sulcus (t. 1) with the last indeed in a three-fold manner.

The *parieto-occipitalis* (po) connects with the *horizontalis* (ho) and has a shallow communication with the *s. temporalis superior* (t. 1) and in an extensive manner with the *scissura hippocampi* (h) and the *s. collateralis* (cl); it penetrates through the fissures of the *praecuneus* (Q), communicates in a shallow manner with the *calloso-marginalis* (cm).

The *calloso-marginalis* itself, as well as the *s. cruciatus* of the *praecuneus*, communicates with the fissure between the *gyrus fornicatus* and the *corpus callosum*.

The *s. temporalis superior* (t. 1) in its anterior part communicates with the *f. Sylvii* and the *interparietalis* (ip). Wernicke's fissure (k) is connected with the upper temporal sulcus (t. 1) and is not clearly separated from the *horizontalis* (ho).

The inferior occipital sulcus (g) as well as Wernicke's fusiform sulcus (t. 3) communicates with the *s. collateralis* (cl), which last extends to the *scissura hippocampi* (h).

The *s. orbitalis* communicates several times with the *fossa Sylvii*. An external orbital sulcus consists of two shallow-connected parts which separate the orbital lobe from both the lower and middle frontal gyri (F. 3, F. 2). There also exists a shallow connection between the external orbital sulcus and the *f. Sylvii*.

OBSERVATION XVII

Antero-posterior chord,	16.4
Hemispheric arch,	26.0
Anterior curve,	15.2
Middle curve,	5.8
Posterior curve,	5.0

SKULL—CONTENTS, 1,330 Ccm.

Horizontal circumference,	51.8
Ear circumference,	31.2
Greatest length,	17.9
Antero-posterior and transverse circum.,	81.5
Greatest breadth,	14.6
Frontal curve,	13.2
Parietal curve,	12.3
Occipital curve,	11.8
Ear base of nose radius,	11.2
Ear-occiput radius,	10.3
Occipital shortening,	10.9
Facial height,	12.1
Frontal height,	7.1
Nasal height,	8.1

No synostoses.

OBSERVATION XVIII.

GAL, Nicolaus; aet. 66, Hungarian; robber and murderer.

LEFT HEMISPHERE.

The upper portion of the *gyrus centralis anterior* (A) deeply divided and the *s. centralis* (c) in connection with the upper frontal sulcus (f. 1).

The *interparietalis* (ip) connects with the *f. Sylvii* (S) and with the upper and middle temporal sulci (t. 1, t. 2).

The *perpendicularis* (po) communicates with the *horizontalis* (ho) and through this with the *interparietalis* (ip) and with the upper and middle temporal sulci (t. 1 and t. 2).

RIGHT HEMISPHERE.

Shallow communication between the *interparietalis* (ip) and the *f. Sylvii* (S) and also between the *parieto-occipitalis* (po) and the *horizontalis* (ho).

(This brain could not be preserved; its description was taken at once upon its arrival.)

OBSERVATION XIX.

EWE, Ludwig, aet. 32, Croatian; professional thief, covetous; gentle, quiet temperament, changeable manner, phlegmatically confessed his crime and died of pleuritis.

Cerebellum well covered by the cerebrum.

LEFT HEMISPHERE.

S. centralis (c) communicates with *f. Sylvii* (S).

The secondary fissure (v) of the upper frontal gyrus (F. 1) very deep.

The inferior sulcus frontalis (f. 2) at the basal border enters a sagittal fissure which separates somewhat distinctly the outer orbital surface of the frontal lobe (F) from the orbital lobe (Ob).

The lower frontal sulcus (f. 1) is also in shallow connection with one of the anterior branches of the *f. Sylvii* (S").

The *interparietalis* (ip) by a connection with the upper depression of the posterior central gyrus (B) is developed into a well marked *retrocentralis*, and it is connected with the *horizontalis* (ho).

The *parieto-occipitalis* (po) is connected with the *horizontalis* (ho) by a deep branch which is given off on the medial surface, and by a curve reaches the external surface. It also communicates with the *scissura hippocampi* (h) and the *s. cruciatus* of the *praecuneus*. Besides this, the *parieto-occipitalis* sends a deep branch from the medial border to the external surface, running anteriorly almost to the upper depression of the *gyrus centralis posterior* (B). The upper temporal sulcus (t. 1) connects with the *f. Sylvii* (S) and extends nearly to the occipital apex and thus the *s. temporalis medius* (t. 2) is separated from the *lobulis tuberis* (P 2') (*gyrus angularis*), and near the occipital apex contains Wernicke's (k) as a radial fissure. This last (k) is connected with the poorly developed lower occipital sulcus (g).

The *lobulis tuberis* (P. 2') (*gyrus angularis*) is divided into two parallel gyri.

The *s. temporalis medius* (t. 2) in this instance represents the *s. fusiformis*.

For this reason, the middle basilar lobe, (U and H) which enters quite abruptly into the posterior basilar lobe, is very broad.

The *s. collateralis* (cl) communicates superficially with the middle temporal sulcus (t. 2) and with the *parieto-occipitalis* (po).

The orbital gyrus (Ob) is extremely broad; the orbital fissure (ob) is very complicated and extends almost to the *fossa Sylvii* (S) at the base. An external orbital sulcus separates the orbital gyrus from the lower frontal gyrus (F. 3).   (See above.)

RIGHT HEMISPHERE.

The *s. centralis* (c) communicates on the surface with the *f. Sylvii* (S) and the upper branch of the *praecentralis*.

The third frontal sulcus (f. 3) by connecting with a vertical branch, which in this case is separated from the upper frontal sulcus (f. 1) forms a *praecentralis*, and it communicates with the *ramus anterior fissurae Sylvii* (S″) and as before said through its upper part, as a *praecentralis*, with the *s. centralis* (c).

In its anterior part the upper frontal sulcus (f. 1) unites with the secondary fissure (φ) whereby the first is, as it were, denied its connection with the vertical branch.

The *interparietalis* (ip) is divided into two parts. It forms a *retrocentralis* and connects with the *ramus posterior fissurae Sylvii* (S′); the upper temporal sulcus (t. 1) and the *occipito-horizontalis* (ho).

The *s. temporalis superior* (t. 1) communicates with the *f.

Sylvii (S), with the *interparietalis* (ip), and indirectly with the *perpendicularis* (po) and *horisontalis* (ho).

The middle and lower temporal sulci (t. 2 and t. 3) together with the corresponding second and third temporal gyri (T. 2, 3) are well developed. The middle basilar lobe (U and H) is very reduced.

The *parieto-occipitalis* (po) is connected by a deep cleft with the *horisontalis* (ho) and this last appears as if it were a prolongation of Wernicke's fissure (k).

The inferior occipital sulcus (g) not distinct.

The *s. collateralis* (cl) is in shallow connection with the *parieto-occipitalis* (po). The *s. cruciatus* of the *praecuneus* sends to the external surface a very deep fissure which reaches almost to the *retrocentralis*. It (cl) is well separated from the *parieto-occipitalis* (po). Respecting its connection with the *calloso-marginalis* (cm), nothing definite can be said for reason that the preparation was damaged on both sides. From the same cause, nothing can be offered concerning the existence or non-existence of an external orbital sulcus upon the right side.

ADDENDA.

OBSERVATION XX.

BOLTER, Male, aet. 43, Crotian; condemned to three years' imprisonment for theft. Great adroitness in contriving rascalities, completely uneducated, unfeeling, rough, obstinate, and in morals thoroughly bad.[2]

Cerebellum completely covered, asymetric. The right cerebellar lobe extended further back, the left one more forward. The left one is imbedded in the deep fossa of the left occipital basilar lobe, whilst the right one lies in a much shallower and shorter niche.

LEFT HEMISPHERE.

S. centralis (c) distinctly separated from the *f. Sylvii* (S) though only by a small bit of gyrus. By a depression of the upper third of the *gyrus centralis posterior* the *s. centralis* is imperfectly separated from the *retrocentralis*.

Separation from the radial branch of the upper frontal sulcus (f. 1) is also indistinct.

The third frontal sulcus (f. 3) connects with the lower (f. 2) and sends a branch anteriorly which connects with the anterior part of the upper frontal sulcus (f. 1) and it is itself

---

[1] After I had concluded the observations (1 to six) I received, through the kindness of Dr. RUBACEK and Director TAUFFER, three more brains, the descriptions of which I here add.

[2] At death, as an "anarchist," he brutally refused the consolation of the priest.

In communication with a little secondary fissure coming from the upper frontal gyrus. The lower frontal sulcus (f. 2) communicates by a shallow fissure with the radial branch of the upper frontal sulcus (f. 1). The upper frontal sulcus has really two radial branches which lie behind one another.

The external orbital fissure is represented by two separate pieces; there is no third branch of the *f. Sylvii*.

The anterior basilar lobe is well developed, with a complicated, though isolated *s. orbitalis cruciatus* (ob).

The *interparietalis* (ip) connects with the *s. cruciatus* of the *praecuneus* (Q) and the *parieto-occipitalis* (po); also with the *ramus posterior fissurae Sylvii* (S') and through this with the upper temporal sulcus (t. 1).

The upper temporal sulcus (t. 1) communicates with the *f. Sylvii*.

The *lobulus tuberis* (P. 2') (*gyrus angularis?*) appears precisely like a connecting curved convolution of the two extended temporal gyri. Wernicke's fissure (k) rises from the *horizontalis* (ho) and communicates with the imperfectly formed *sulcus occipitalis inferior*. An anterior, isolated piece of the *sulcus occipitalis inferior* is represented by the posterior end of the *fissura fusiformis*, of which we will soon speak. The third temporal sulcus (t. 3) extends to the posterior pole and is composed of two parts which have a shallow communition with each other. The anterior part of it runs chiefly along the base and communicates with the *s. collateralis* (cl).

The middle basilar lobe is very undeveloped. In its anterior portion it appears like a loop of the middle and lower (T. 2, T. 3) temporal gyri, and in its posterior part as a loop of the extraordinarily long *gyri lingualis* and *fusiformis*.

The *parieto-occipitalis* (po) communicates with the *horizontalis* (ho) and also with the *interparietalis* (ip) and Wernicke's fissure (k). The *cuneus* is very greatly dwarfed so that on the medial surface the *gyrus lingualis* shows a greater height than the *cuneus*.

The *calloso-marginalis* (cm) in its anterior portion is composed of two pieces. It is well isolated from the fissure system of the *cuneus*, but by a shallow, lateral continuation penetrates far into the external surface.

The *S. cruciatus* of the *praecuneus* communicates, as before
said, with the *interparietalis* (ip), and also connects with the
fissure between the *gyrus fornicatus* and *corpus callosum* and
through the shallow bed of a vessel, a communication with
the *parieto-occipitalis* (po) is indicated.

Antero-posterior chord,	17.4
Hemispheric arch,	22.0
Anterior curve,	13.0
Middle curve,	3.0
Posterior curve,	6.0

RIGHT HEMISPHERE.

S. centralis (c) not separated from the *f. Sylvii*. Third
frontal sulcus (f. 3) has an extremely shallow communication
with the *ramus anterior fissurae Sylvii* (S''). The *praecentralis*
is composed of three pieces, the lower one of which
corresponds with the *s. frontalis tertius* (f. 3), the middle to a
separated radial branch of the upper frontal sulcus (f. 1), and
the upper one to a separated branch of the secondary fissure
(g). The upper frontal sulcus has besides this another large
radial branch.

The second anterior incision of the *fossa Sylvii* forms a
very arborescent *s. orbitalis cruciatus* within the lower frontal
gyrus, and communicates with the inferior frontal sulcus (f. 2).

A third off-shoot from the *fossa Sylvii* is only indicated by
the bed of a vessel. The middle portion of the external
orbital fissure is well developed, as is also the anterior portion,
and its connection with the middle part is only indicated.

Orbital lobe well developed; its *s. cruciatus* isolated.

The *interparietalis* (ip) is fully developed into a *retrocentralis*,
and communicates with the *horizontalis* (ho), and *parieto-occipitalis* (po),
the upper temporal sulcus (t. 1), and through
that with the *f. Sylvii* (S).

The upper temporal sulcus (t. 1) communicates with the
f. Sylvii, the *interparietalis* (ip), and also by very shallow
fissures with the occipital fissure.

Wernicke's fissure (k) also communicates with the occipital fissure.
The *s. occipital. inferior* well developed and in
connection with upper temporal sulcus (t. 1).

Fissura fusiformis poorly developed, and in connection with upper temporal sulcus (t. 1). *Gyri lingualis* and *fusiformis* extremely dwarfed and also the middle basilar lobe.

S. collateralis extends closely to the *scissura hippocampi* (h) and communicates with the *fissura fusiformis*.

Fissura calcarina (cc) separate from the *parieto-occipitalis* (po); *cuneus* very strongly developed.

The *parieto-occipitalis* (po) communicates with the *horisontalis* (ho) and the *interparietalis* (ip) and through these with the upper temporal sulcus (t. 1) and Wernicke's fissure (k).

The *calloso-marginalis* (cm) isolated. The same is true of the *s. cruciatus* of the praecuneus, the lowest sagittal fissure of which makes an extensive separation between the *praecuneus* and the *gyrus fornicatus*.

Both hemispheres are extraordinarily long and low.

Antero-posterior chord,	17.2
Hemispheric arch,	22.0
Anterior curve,	12.6
Middle curve,	3.4
Posterior curve,	6.0

This brain also belongs to the type of "confluent fissures," and the separation of the *parieto-occipitalis* (po) from the *fissura calcarina* (cc) on the right, is to be considered as especially atypical (see Obs. V.) The cranium belonging to the brain was evidently in a high degree dolichocephalic.

OBSERVATION XXI.

KRISTIC, Jovo, æt. 26; Croatian. First punishment for theft. Intellectually rather capable, uneducated, rough, frivolous, drunkard. In prison, obedient.

Cerebellum poorly covered on the left and extremely uncovered on the right; middle portion uncovered. Right side of cerebellum extended much further back than the left.

LEFT HEMISPHERE.

S. centralis (c) in communication with the *f. Sylvii* by a shallow fissure which penetrates through the lower part of the *gyrus centralis superior* (B). There is also a very shallow communication with the radial branch of the upper frontal sulcus. The lower part of the posterior central gyrus is divided by a fissure which has free connection with the *f. Sylvii* and a shallow one with *interparietalis* (ip); it, however, does not extend to the *s. centralis* (c).

The third frontal sulcus communicates with the *ramus anterior fissurae Sylvii* (S″) but not with the Inferior frontal sulcus (f. 3).

The *praecentralis* is composed of three parts, the lower one of which (f. 3) communicates with the short radial branch of the upper frontal sulcus, whilst the upper parts pertain to the secondary fissure (φ) of the upper frontal gyrus.

Middle frontal gyrus (F. 2) very undeveloped.

Upper frontal gyrus (F. 1) exhibits deep secondary fissures.

A third incision of the *fossa Sylvii* is well developed. The middle and anterior parts of the external orbital fissure present, and also three isolated from each other.

The orbital lobe (Ob) is rather broad but short; the orbital sulcus (ob) well developed.

The *interparietalis* is developed into a complete *retrocentralis*. At its lower end it is imperfectly separated from the *f. Sylvii*, and it connects with it by the previously mentioned fissure of the lower portion of the posterior central gyrus (B). It connects with the *horizontalis* (ho), is ill-separated from the *perpendicularis* (po) and on the other hand extends far into the *praecuneus*.

The *s. temporalis superior* (t. 1) communicates with the *f. Sylvii* (S) and with its posterior ramus (S').

Wernicke's fissure (k) and the lower occipital fissure (g) poorly characterized. The middle temporal sulcus (t. 2) extends far back and communicates with the upper temporal sulcus (L 1).

The *fissura fusiformis* runs forward entirely along the base where it communicates with the *s. collateralis*. The *sulci collateralis* and *fusiformis* united, penetrate the middle basilar lobe parallel with its medial border, and not more than 2–4 Cm. distant from it, therefore, the occipital portion of the middle basilar lobe (U and H) form, at the under surface, only a thin, short strip. The entire *gyrus hippocampi* and a part of the *gyrus uncinatus* really lie at the medial surface, whilst the lower surface of the middle basilar lobe is at most exclusively occupied by the middle and inferior temporal gyri (T. 2, T. 3). A fissure extending back from the *f. Sylvii* separates the *gyrus hippocampi* from the *gyrus uncinatus*.

Gyrus fusiformis to a great extent, undeveloped; the same with the anterior portion of the *gyrus lingualis*.

The *parieto-occipitalis* (po) communicates, as before mentioned, with the *interparietalis* (ip), and also with the *scissura hippocampi*, and by means of the fissures of the *praecuneus* with the *calloso-marginalis* (cm). This (cm) has an independent anterior part which sends a short branch to the medial edge (analogous to the *fissura cruciata* of Leuret) commencing in the praecentral region it penetrates through the *parieto-occipitalis* (po) to the *scissura hippocampi*.

Antero-posterior chord,	16.5
Hemispheric arch,	22.0
Anterior curve,	13.0
Middle curve,	4.0
Posterior curve,	5.0

RIGHT HEMISPHERE.

S. centralis (c) communicates with *f. Sylvii* (S).

The third frontal sulcus (f. 3) connects with the anterior ascending ramus of the *fissura Sylvii* (S") and communicates with the lower frontal sulcus (f. 2) which in its turn communicates with the upper frontal sulcus (f. 1). The lowest part of the *praecentralis* (f. 3) is separate from the upper. That above has two parts which lie behind one another and which communicate with each other and with the upper frontal sulcus (f. 1).

The last one is peculiarly constructed in that its posterior part ascends and really represent a secondary fissure in the upper frontal gyrus. The other part of the upper frontal gyrus (F. 1) contains a deep secondary fissure.

A third incision of the *fossa Sylvii* penetrates about 2.5 Cm. towards the frontal extremity. Parallel with and above this, there is a middle part of the external orbital fissure which, by an independent, shallow fissure, connects with the *f. Sylvii*. The isolated anterior portion of the external orbital fissure is but little developed.

The orbital lobe is short and the orbital fissure (ob) extends into the *f. Sylvii*.

The *interparietalis* is connected with the *f. Sylvii*, is developed into a *retrocentralis* and communicates with the *horisontalis* (ho) and *perpendicularis* (po).

The upper temporal sulcus (t. 1) communicates with the *f. Sylvii*. The upper and middle temporal gyri extend very far forward. The upper temporal sulcus (t. 1) communicate with Wernicke's fissure and the *s. occipitalis inferior* (t. 3).

The occipital lobe behind the *horisontalis* (ho) and Wernicke's fissure, lies more upon the posterior than the external lateral surface. The *perpendicularis* (po) as before said, communicates with the *horizontalis* (ho) and the *interparietalis* (ip) and through the praecuneal fissure with the *calloso-marginalis* (cm) and also in the most extensive manner with the *scissura hippocampi*.

The *gyrus lingualis* does not lie on the base but on the medial surface, and here again the innermost very small part of the middle basilar lobe appears as a loop of the two gyri of the undeveloped occipital basilar lobes. The *fissura fusiformis* (f. 3) lies entirely at the base, nearer to the medial than outer edge, whereby only a small strip 1.5 Cm. broad remains on the inner border of the middle basilar lobe for the *gyrus uncinatus*, whilst the *gyrus hippocampi* lies chiefly upon the inner surface. The *gyri uncinatus* and *hippocampi* are here again separated by a fissure emerging from the *fossa Sylvii*.

From what has been said, the immense development of the temporal lobe in this brain becomes evident.

The *calloso-marginalis* (cm) forms a complete curve from the lowest part of the medial frontal lobe to the *scissura hippocampi*, from which, in the frontal wedge, arises a condition analogous to the *fissura cruciata of Leuret*; on this side analogous to the horse's brain, whilst on the other side it corresponds more to the bear's brain.

Antero-posterior chord,	15.7
Horizontal arch,	22.0
Anterior curve,	10.3
Middle curve,	8.2
Posterior curve,	3.5

This brain also belongs to the *confluent fissure* type and it is seen to be especially atypic in the extreme degree to which the *gyri uncinatus* and *hippocampi* are dwarfed. They represent only a puny, pushed-backwards-and-inwards, portion of the middle basilar lobe. From this peculiarity these two hemispheres could be detected amongst thousands of others.

OBSERVATION XXII.

ZATEZALO, Vaso, aet. 25, Croatian. Had been previously punished for dangerous threats. Was at this time condemned to five years' imprisonment for manslaughter; he ripped open his antagonist's abdomen. The man was addicted to drink, had weak mental powers, and possessed the lowest grade of cultivation.

The cerebellum was almost entirely uncovered.

Both hemispheres highly hypsocephalic upon a short base, and fissured in a manner reminding one more of a cetaceous than a human brain.

LEFT HEMISPHERE.

S. centralis (c) communicates with *praecentralis* and in a very shallow manner with the *interparietalis* (ip) and by a fissure which penetrates through the lowest part of the posterior central gyrus (B) with the *f. Sylvii*, with which it also communicates indirectly by way of the *interparietalis*. The radial branches of the *calloso-marginalis* (cm) are so multiple and irregular that by themselves they offer no reliable guide for locating the *sulcus centralis*.

The *s. centralis* (c) runs so diagonally that the lower end is 6 Cm. further forward than the upper end.

The *praecentralis* runs entirely parallel with the *s. centralis* and sends a branch bending upwards and forwards, thereby forming a sagittal branch to a secondary fissure (v).

The *praecentralis* communicates with the *fossa Sylvii* and as before said, with the *s. centralis*.

The upper frontal gyrus lies chiefly upon the medial surface and is separated in its entire extent from the *gyrus fornicatus* by a well-defined upper *calloso-marginalis* (cm).

The *s. frontalis inferior* (f. 2) also communicates with the *fossa Sylvii*.

The unusually deep external orbital fissure which here separates the middle and lower frontal gyri from the orbital lobe is of especial interest. It does not stand really in direct connection with the *fossa Sylvii*; but as the middle basilar lobe in this case extends far in front of the anterior end of the lower frontal gyrus, it results that the posterior end of this orbital fissure is covered by the temporal gyri.

The *lobus olfactorius* has an uncommonly strong development.

The *gyrus orbitalis* points partly outwards instead of downwards.

The *interparietalis* (ip) communicates, as before said, with the *f. Sylvii*. It is in communication with the *horizontalis* (ho) and the *perpendicularis* (po) and also with the posterior ramus of the *f. Sylvii* and with the *s. temporalis superior* (t. 1).

The upper temporal sulcus (t. 1) has several communications with the *f. Sylvii* and also as before said, with the *interparietalis* and through this with the occipital fissures.

Wernicke's fissure (k) connects with both the *interparietalis* and upper temporalis (t. 1) and can be traced to the lower border of the external surface. It intersects a transverse fissure which must be regarded as the *s. occipitalis inferior* and through that communicates with the middle temporalis (t. 2) and further up with the first temporalis (t. 1).

The *perpendicularis* (po) as mentioned, communicates with the *interparietalis*, the *horizontalis*, the upper temporalis and Wernicke's fissure.

It is clearly separated from the *scissura hippocampi* and also from the fissures of the *praecuneus*.

The *gyrus lingualis* lies mostly on the medial surface. The posterior part of the dwarfed middle basilar lobe affords the connection between the two gyri of the occipital basilar lobe, whilst the anterior part of the dwarfed middle basilar lobe connects with the temporal lobe.

The *fissura fusiformis* is composed of two separate parts; the anterior one of which separates the temporal from the middle basilar lobe. The posterior one separates the *gyrus fusiformis* from the temporal lobe, and communicates with the temporal sulci, Wernicke's fissure and the *s. occipitalis inferior.* The deepest situated part of the brain is the occipital basilar lobe. The middle as well as the anterior basilar lobes are placed higher; as seen from beneath, the posterior part of the brain stands higher than the middle and anterior.

The *calloso-marginalis* is greatly branched. It sends several deep radial branches towards the medial border and communicates with the fissure of the *praecuneus.* The latter, isolated externally and posteriorly, sends several shallow branches to the fissure between the *gyrus fornicatus* and the *corpus callosum.* The *gyrus fornicatus* is extremely short, more circular than oval, rolled over the spherical triangle of the *thalamus opticus.*

Antero-posterior chord,	15.51
Hemispheric arch,	24.31
Anterior curve,	14.3
Middle curve,	5.7
Posterior curve,	4.3
Greatest height,	8.9

RIGHT HEMISPHERE.

The *s. centralis* (c) communicates with the *f. Sylvii* by a fissure which passes through the lower part of the posterior central gyrus and which, towards its terminus, is very shallow. It also communicates with the upper frontal sulcus (f. 1). There exists a *praecentralis* which extends from the *fossa Sylvii* quite into the medial border and its upper part only stands in communication with a sagittal fissure. Besides this, there is a second *praecentralis*, the lower part of which is in connnection with the middle frontal sulcus (f. 2). Its upper part communicates with no sagittal fissure, and represents a separated radial fissure from the upper frontal sulcus (f. 1).

Upper frontal sulcus extremely undeveloped, whilst the secondary fissure (φ) extends far forwards from the *s. centralis.*

The lower and middle frontal gyri much dwarfed. The

anterior part of the frontal lobe on this side rises almost vertically, whilst, although steep on the other (left) side, still it runs somewhat pointedly forwards.

There is a third sylvian incision but very shallow. The middle and anterior parts of the external orbital fissure unite with each other and are well developed.

The orbital lobe lies at the base, but is extraordinarily dwarfed, whilst the olfactory lobe lies partly on the medial surface and is very largely developed.

The *interparietalis* (ip) consists of 1st, a fissure rising from the *f. Sylvii*, running quite parallel with the *s. centralis* and having no connection with its sagittal part: 2d, the sagittal part communicates with the *horisontalis* (ho); upper temporal sulcus (t. 1); also with the fissure system of the *praecuneus* and through this with the *calloso-marginalis* (c).

The *s. temporalis superior* (t. 1) is in two parts, the upper one of which, as before stated, communicates with the *interparietalis* (ip) and in this way with the *horisontalis* (ho). The lower part is much dwarfed. Wernicke's fissure is well developed.

The *parieto-occipitalis* (po) is indistinctly separated from the *horisontalis* (ho) and blends indistinctly with the *scissura hippocampi*.

Gyrus lingualis thick and short; and entire occipital basilar lobe cut up into islands. The *s. collateralis* communicates anteriorly with the *fissura fusiformis*.

The *collateralis* (cl) continues across the posterior border of the brain to the anterior and external surface and extends nearly to the posterior branch of the *fissura fusiformis*, connects with the middle temporal sulcus (t. 2).

The middle and lower temporal gyri (T. 2, T. 3) well developed, but that part of the middle basilar lobe which connects with the two gyri of the occipital basilar lobe is in its temporal part dwarfed.

The *gyrus fusiformis* is, by a transverse fissure, completely separated from the middle basilar lobe.

The *calloso-marginalis* (cm) in its anterior portion sends a branch to the external surface (analogous to the *fissura cruciata of Leuret*) and through the fissure system of the *praecuneus* extends nearly to the *scissura hippocampi*.

The fissure system of the *praecuneus* communicates with the fissure between the *corpus callosum* and the *gyrus fornicatus*.

The *Cuneus* is not towards the medial surface but towards the posterior. The occipital extremity points entirely downwards and outwards.

Antero-posterior chord,	14.9!
Hemispheric arch,	24.81
Anterior curve,	13.3
Middle curve,	5.4
Posterior curve,	6.1
Greatest height,	9.9

I conclude with perhaps the most atypic brain of those we have hitherto examined. The cranium was evidently a towarskull.

Upon comparing the last three brains, it is to be seen that the measurements of the base chord and the hemispheric curve, even with all their defects, are yet very instructive.

The brain of Obs. XXI. shows the average proportion of 1: 1.37, and evidently most nearly corresponds to the normal proportions of the south Slavic brains.

The brain of Obs. XX. exhibits the average proportion of 1: 1.27. This brain was flat and long.

The third brain (Obs. XXII) was in a high degree hypsocephalic and the average proportion of the base-chord to the hemispheric curve is 1: 1.61!

RECAPITULATION I.

(EPILEGOMENA).

I.

First, let us consider, statistically, the connections of typical fissures.

I. That of the PARIETO-OCCIPITALIS (po) with the *horisontalis* (ho) and *sulcus interparietalis* (ip).

Upon 38 hemispheres:

It is complete—On the right,	12 times.
On the left,	9 "
Total	21 "
Incomplete (that is by shallow fissures):	
On the right,	3 times.
On the left,	3 "
Total,	6 "
Connection absent:	
On the right,	4 times.
On the left,	7 "
Total,	11 "

Therefore, in the 38 cerebral hemispheres, the *parieto-occipitalis* (po) unites with the *horisontalis* (ho) 27 times.

Ten times it was on both sides: 5 times it was on the right side, and twice on the left. There remain eleven cerebral hemispheres without this connection, but there are only two cases where it is absent in both sides. In three of the eleven hemispheres the separation is poor.

Of 19 brains, therefore, there are only two in which the communication is lacking upon both sides.

II. On account of its analogy with the *"ape-fissure,"* it is interesting to study the connections of the united PARIETO-OCCIPITALIS (po) and HORIZONTALIS (ho).

(a) WITH WERNICKE'S FISSURE (K).

It is complete on the right,	3 times.
It is complete on the left,	1 "
Incomplete on the right,	1 "
" " " left,	1 "
Total,	6 "

In six of the cerebral hemispheres there is a communication completely analagous to the *"ape-fissure."* It exists on both sides in only one brain. In one brain the data on this point are lacking.

(b) The united *perpendicularis* (po) and *horizontalis* (ho) blend with the first or second temporal sulci (t. 1 and t. 2) (and indeed to a certain degree indirectly so through the *sulcus interparietalis* (ip)):

Complete on the right,	7 times.
Complete on the left,	6 "
Incomplete on the right,	2 "
Incomplete on the left,	1 "

Concerning two hemispheres no statement.

In twenty-five hemispheres there are sixteen connections between the united *perpendicularis* (po) and *horizontalis* (ho), and the temporal sulci. Four times in six, the connection occurs as under (*β*) united with that described under (*α*). Thus there is a connection in 18 hemispheres.

The blending of the *perpendicularis* (po) and *horizontalis* (ho) predispose to a union of these with Wernicke's fissure or with the temporal sulci.

III. The HORIZONTALIS (ho) may, however, be separate from the *perpendicularis* (po) but united to Wernicke's fissure or to the temporal sulci, which can hardly be considered as coming within the normal type.

We will now consider eleven hemispheres, in which there is no union of the two occipital fissures (po and ho).

Of these, one is excluded on account of injury to the preparation.

Of the remaining ten there can be shown a complete direct or indirect connection in . . . . . 2 cases.
Incomplete in . . . . . . 4 "
_____
Total, ... 6 "

If we combine the results given under 2 and 3, it will be found that in 35 cerebral hemispheres the *horizontalis* (ho) communicates with either Wernicke's fissure or the temporal sulci, not less than 24 times.

IV. In the next place the relation of the SULCUS CENTRALIS to the surrounding fissures is interesting. We will first consider:

(*a*) the connection of the *sulcus centralis* with the *fissura Sylvii* (S).

Of 38 cerebral hemispheres this union is

Complete on the right, ... 9 times.
Complete on the left, ... 9 "
_____
Total, ... 18 "
Incomplete (by shallow fissures) on the right, 2 "
Incomplete " on the left, 4 "
_____
Total, ... 6 "

Thus, of thirty-eight (38) cerebral hemispheres, there are 24 in which the *sulcus centralis* is not separate from the *fissura Sylvii.* This does not include the indirect communications by means of the frontal and interparietal sulci.

This connection occurs on both sides in 9 cases out of 19, and on one side alone in six cases; there being no connection in four only of the 19 brains.

(β) The *sulcus centralis* communicated with the frontal sulci.

RECAPITULATION L

I. With the third frontal sulcus (f. 3):

Completely on the right,	4 times.
Completely on the left,	7 "
Total,	11 "
Incompletely on the right,	2 "
Incompletely on the left,	0 "
Total,	2 "

13 times in 38 hemispheres.

In nine brains the communication was absent on both sides, in seven, upon one side. Communication occurred on both sides only in three cases.

II. The *sulcus centralis* (c) communicated with the *sulcus frontalis superior* (f. 1).

Completely on the right,	1 times.
Completely on the left,	8 "
Total,	9 "
Incompletely on the right,	1 "
Incompletely on the left,	0 "
Total,	1 "

The *sulcus centralis* in no instance communicated with the *sulcus frontalis superior* on both sides.

Of the 19 brains there are only three (6 hemispheres) in which there is no communication on either side between the *sulcus centralis* and a frontal sulcus (KUSS, SZINKA, PROKETZ).

Only in four brains was the *sulcus centralis* connected with the upper frontal sulcus (f.1) and not with the third frontal sulcus (f. 3). In six brains it was connected with both the upper and the third frontal sulci, and this was in all cases only on one side.

Of the three brains in which there was no communication between the *sulcus centralis* and the frontal fissures, there were two in which the *sulcus centralis* on both sides connected with the *fissura Sylvii* (SZINKA, KUSS).

(γ) There was communication between the *s. centralis* and *s. interparietalis.*

Complete on the right,	2 times.
Complete on the left,	5 "
Total,	7 "
Incomplete on the right,	2 "
Incomplete on the left,	2 "
Total,	4 "

In only one brain is the connection on both sides, so that the communication exists in ten brains. Amongst these is the brain of PROXETZ.

Of the 19 brains there is not a single one in which the *sulcus centralis* has not, at least on one side, a connection with some other fissures.

Altogether there are 58 connections, of which 35 are on the left and 23 upon the right side.

In five of the brains there were connections only upon one side, and this in each case was upon the left side; In three of the five there existed only one connection in each. In only one instance did the connections on the right outnumber those on the left side.

V. We will now examine the communications of the FISSURA SYLVII.

(α) With the *sulcus centralis* (see above) the connection exists 24 times, to wit:

Completely on the right,	9 times.
Completely on the left,	9 "
Total,	18 "
Incompletely on the right,	2 "
Incompletely on the left,	4 "
Total,	6 "

(β) With the frontal sulci:

Completely on the right, ...	9 times.
Completely on the left, ...	9 "
Total, ...	18
Incompletely on the right, . . .	3 "
Incompletely on the left, ...	4 "
Total, ...	7 "

In 25 of the 38 brains there existed a connection between the *fissura Sylvii* (its anterior ramus) and the frontal sulci.

Connection is absent on both sides in only three brains (six hemispheres); in seven brains its exists on both sides.

But in those instances where the direct communications failed, it was effected by a communication of the third frontal sulcus with the *sulcus centralis*, and this last with the *fissura Sylvii*.

These indirect communications were:

On the right, ...	1
On the left, ...	1,

so that in 27 of the 38 hemispheres the *fissuras Sylvii* could be entered from the third frontal sulcus.

(γ) A third communication between the interparietalis and *fissura Sylvii* existed.

Complete on the right, ...	12 times.
Complete on the left, ...	10 "
Total, ...	22 "
Incomplete on the right, ...	4 "
Incomplete on the left, ...	2 "
Total, ...	6 "

In 28 of the 38 hemispheres then, there existed a direct connection between the *interparietalis* and *fissura Sylvii* (either posterior ramus or horizontal part).

(?) Communication between the *fissura Sylvii* and *sulcus temporalis superior* (t. 1) existed

Complete on the right, ...	10 times.
Complete on the left, ...	8 "
Total, ...	18 "
Incomplete on the right, ...	2 "
Incomplete on the left, ...	2 "
Total, ...	4 "

The *fossa Sylvii*, in the 38 cerebral hemispheres, was in direct communication with the *sulcus temporalis superior* (t. 1) twenty-two times.

The *fossa Sylvii* communicated extensively with the *fissura orbitalis* (ob).

On the right, ...	7 times.
On the left, ...	7 "
Total, ...	14 "

In six of these fourteen, the communication was upon both sides. This connection existed, therefore, only in eight brains. Four cerebral hemispheres could not be investigated; so that in 34 hemispheres there were 14 connections between the *fissura orbitalis* and *fossa Sylvii*.

In 38 cerebral hemispheres we have no less than 1131 connections between the *fissura Sylvii* and the surrounding fissures, and with some of them, moreover, the communication is repeated.

VI. The SULCUS TEMPORALIS SUPERIOR (t. 1) is in connection with:

(a) The *fissura Sylvii* (horizontal part and posterior ramus, see under 5, *b*).

Completely, ...	18 times.
Incompletely, ...	4 "
Total, ...	22 "

With the *sulcus interparietalis* (ip):

Completely on the right, ...	10	times.
Completely on the left, ...	9	"
Total, ...	19	"
Incompletely on the right, . . .	4	"
Incompletely on the left, ...	2	"
Total, ...	6	"

The upper temporal sulcus (t. 1) joined with the *sulcus interparietalis* (ip) in all 25 times.

This connection was absent in 13 hemispheres, but in no single brain was it lacking upon both sides. In 7 brains it was present on both sides. In those hemispheres where these two sulci were not connected, connections between the temporal sulci and *horizontalis* (ho) were exceptional.

The upper temporal sulcus (t. 1) then, in 38 cerebral hemispheres, was connected completely 37 times with the *fissura Sylvii* (S and S'), and with the *sulcus interparietalis*, (ip,) in a shallow manner, ten times, in all 47 times.

Besides this there exist connections with Wernicke's two fissures, which cannot be regarded as atypical, also with the *fissura fusiformis* and through it with the *sulcus collateralis* (cl). An extensive communication with the fissures of the base might be viewed as atypic.

As belonging to the atypical connections of the *sulcus temporalis superior* (t. 1) might be reckoned a connection with the *horizontalis* (ho) or the conjoined *horizontalis* (ho) and *parieto-occipitalis* (po).

This connection exists (direct or indirect):

Complete on the right, ...	9	times.
Complete on the left, ...	6	"
Total, ...	15	"
Incomplete on the right, ...	4	"
Incomplete on the left, ...	3	"
Total, ...	7	"

Through injury to two of the hemispheres, the connection in them cannot be given.

In 22 of the 36 hemispheres, the upper temporal sulcus extends into the upper occipital sulcus.

With the fissures mentioned under α, β, and γ, the upper temporal fissure has 69 connections.

VII. We will now consider the connections of the SULCUS INTERPARIETALIS (ip).

(α) With the sulcus centralis (c) (see above, under 4 γ),

Complete, ...	7 times.
Incomplete, ...	4 "
Total, ...	11 "

(β) With the *fissura Sylvii* (see above under 5 γ).

Complete, ...	22 times.
Incomplete, ...	6 "
Total, ...	28 "

(γ) With *sulcus temporalis superior* (t. 1). (See under 6 β).

Complete, ...	19 times.
Incomplete, ...	6 "
Total, ...	25 "

A rare but certainly interesting connection is:

(δ) The *interparietalis* (ip) with the *sulcus calloso-marginalis* (cm) or with the *sulcus cruciatus* of the *praecuneus*.

This was found three times, twice on the right (Obs. II and X) and once on the left (Obs. IV).

In 38 cerebral hemispheres then, we have 51 complete and 16 shallow, in all 67 connections of the *interparietalis*.

In no brain was there an absence of the connection on both sides; in two of the brains there were, including both sides, no less than five communications.

VIII. We will now observe the undoubtedly atypical connection of the SCISSURA HIPPOCAMPI (h, of Fig. II of the Introduction). First:

(α) With the *fissura parieto-occipitalis* (po),
The communication was

Complete on the right, ...	10 times.
Complete on the left, ...	7 "
Total, ...	17 "
Incomplete on the right, ...	1 "
Incomplete on the left, ...	1 "
Total, ...	2 "

The connection of the *scissura hippocampi* (h) with the *parieto-occipitalis* (po) was complete 17 times and incomplete 2 times, in all, 19 times. This included 14 brains out of the 19. The commuications existed on both sides, however, in only five brains.

(β) The *sulcus collateralis* (cl) communicated with the *scissura hippocampi* (h),

Complete on the right, ...	5 times.
Complete on the left, ...	4 "
Total, ...	9 "
Incomplete on the right, ...	2 "
Incomplete on the left, ...	0 "
Total, ...	2 "

It is thus seen that the *sulcus collateralis* communicates 11 times with the *scissura hippocampi* (h) and this almost without exception, in those cerebral hemispheres where there was also a communication with the *parieto-occipitalis* (po).

(γ) The sulcus *calloso-marginalis* (cm) communicated but once, directly, with the *scissura hippocampi* (h), (Obs. VII) though often communicating by way of the *parieto-occipitalis* (po). Between other fissures and the *scissura hippocampi* there were 31 (11) connections.

IX. A study of the fissure-system of the SULCUS CALLOSO-MARGINALIS (cm) is highly important.

As concerns the human brain, it is accepted as typical that

this sulcus should rise up from the frontal lobe and, with a posteriorly convex curve, find its way to the external surface back of the posterior central gyrus (B). It often sends an anteriorly convex fissure which defines the anterior border of the *gyrus centralis anterior* (A). More frequently, however, this last mentioned curve exists unconnected with the *sulcus calloso-marginalis*, and now and then is absent altogether.

Still more frequently there is a second *sulcus calloso-marginalis* (See Figs. III in Tables I, II, IV, VII, IX, X), in which case the anterior one extends further down towards the base of the medial frontal lobe, whilst the posterior one is unusually short in its front extremity. This last generally furnishes the curved fissure which defines the posterior limit of the paracentral lobe, whilst the first one exhibits a tendency to extend to the external surface as the anterior boundary of the paracentral lobe (See for example Fig. III, Tables I and IX) (a few further modifications exist in ZERNOFF).

This fissure bounding the anterior limit of the paracentral lobe has its analogue, for example, in the bear, where its incision upon the surface forms the *sulcus cruciatus* of Leuret (See I, Fig. III of Recapitulation) and is separate from the *sulcus calloso-marginalis* (cm).

It is regarded as typical that the *sulcus calloso-marginalis* should extend no further backwards.

The simplest form of a continued extension of this sulcus is, that, after the curve which defines the posterior border of the paracentral lobe, the sulcus (cm) is prolonged posteriorly, furnishing a dividing line between the *praecuneus* (Q) and the *gyrus fornicatus* (Gf). (See Fig. III, Pl. XII).

A further combination is a union of this last-mentioned prolongation, directly with the *parieto-occipitalis* (po) or indirectly with it by the superimposed fissures of the *praecuneus* (see for example Figs III in Tables II & XI). It may also connect with the fissures of the *praecuneus* whilst these do not connect with the *parieto-occipitalis* (See Fig. III in Tables I, IV, VII, IX), or, the *calloso-marginalis* (cm) may be separate from the praecuneal fissures and these be in communication with the *parieto-occipitalis* (po). (See Fig. III in Tables V and VI).

These communications are of great interest on account of their relations to comparative anatomy.

In beasts of prey, for example, the *sulcus calloso-marginalis* extends from behind forwards nearly and sometimes unites to the *sulcus cruciatus* of Leuret, or, with an anteriorly convexed curve, it extends almost, or entirely, to the upper medial border. Posteriorly this curve partially encircles the occipital end of the *gyrus fornicatus* (Gf) and this part of it represents the common stem of the *fissurae parieto-occipitalis* (po) and *calcarina* (cc) (in man).

Whilst with the primates the typical *fissura callosomarginalis* separates only the frontal and central portions of the hemisphere from the *gyrus fornicatus*, we see that in our specimens the parietal and even the occipital portions are separated from the *gyrus fornicatus* by it.[1]

We will give a numerical representation of these similarities to animal anatomy, omitting the simple extension of the *sulcus calloso-marginalis*, by which the *praecuneus* is separated from the *gyrus fornicatus* (Gf) and also the direct communication with the *scissura hippocampi* (h).

(α) Communication of *sulcus calloso-marginalis* with *fissura parieto-occipitalis* (po).

Complete on right,	5 times.
Complete on left,	3 "
Total,	8 "
Incomplete on right,	1 "

Thus, in 33 hemispheres there were 9 complete connections. (In 5 of the hemispheres no observations were made).

(β) Communications of the *calloso-marginalis* exclusively with the praecuneal fissures.

---

[1] A communication of the *sulcus calloso-marginalis* or even of the praecuneal fissures with the fossa between the *gyrus fornicatus* and *corpus callosum* (CC) is an exception (peculiarity) (See Fig. III, Tables I, VII, XII). This connection is rarely observed except with the *scissura hippocampi* (h) itself.

RECAPITULATION I.

Complete on the right, ...	2 times.
Complete on the left, ...	5 "
Total, ...	7 "
Incomplete on the left, ...	1 "

Thus there are eight communications between the *sulcus calloso-marginalis* and the praecuneal fissures.

(γ) Communication of the *parieto-occipitalis* (po) with the anteriorly isolated fissures of the *praecuneus*,

Complete on the right, ...	3 times.
Complete on the left, ...	3 "
Total, ...	6 "
Incomplete on the right, ...	1 "
Incomplete on the left, ...	1 "
Total, ...	2 "

Thus there were also eight communications between the *parieto-occipitalis* and praecuneal fissure.

X. The communications of the SULCUS COLLATERALIS (cl) are also certainly important as concerns comparative anatomy.

The fissure which divides the *gyrus uncinatus* (U) from the hippocampus (H) will be omitted, for, as it is a branch of the *sulcus collateralis*, it is perhaps typical.

It is interesting, however, that in one instance this fissure connected with the *sulcus temporalis superior* (t. 1). (See Obs. X).

We will first consider then (a) the communication between the *s. collateralis* (cl) and the *sulci temporales* (t. 1, t. 2, t. 3).

It was extensive on the right, . . .	3 times.
" " " " " left, . . .	2 "
Total, ...	5 "
It was shallow on the left, . . .	2 "

In all, the *collateralis* communicated seven times with the temporal sulci.

(β) It communicated with *parieto-occipitalis* (po) or the *f. calcarina*;

Extensively on the right,	3 times.
" " " left,	3 "
Total,	6 "
It was shallow on the right,	1 "
" " " " " left,	2 "
Total,	3 "

(γ) On each side there were two full communications with the *sulcus occipitalis inferior* (g), four in all.

As we must exclude from consideration Obs. XVIII on account of incomplete examination, we have then 20 communications in 36 cerebral hemispheres. Amongst them were six brains with bilateral communications, and in two hemispheres there were double communications. There were 18 hemispheres (12 brains) which exhibited this connection.

(δ) See under 8 β). There are to be added eleven communications between the *sulcus collateralis* and *scissura hippocampi* (h), three of which, however, are effected indirectly through the *parieto-occipitalis*, so that this only increases by eight the number of the collateralis connections.

In 36 cerebral hemispheres then there were 28 communications.

X. The following were the conditions respecting the COVERING OF THE CEREBELLUM BY THE OCCIPITAL LOBES. With one brain (Obs. XVIII) there was no accompanying statement. With two of the brains (Obs. XIII and XIV), the cerebellum, on account of the flatness of the occipital basilar lobes, was strongly dipped downwards.

In the sixteen remaining brains the covering was:

Extensive,	4 times.
Barely sufficient,	3 "
Insufficient,	3 " !
In great measure wanting,	6 " !!

I will here make a remark respecting the fissure which

I have designated as the "*external orbital fissure.*" We shall see further on that this fissure is very much developed in the gyrocephalic mammalia, which rank beneath the primates.

So far as I know, it is also very constant in apes.

It is not distinct in the human brain, and my attention was directed to it by Obs. II (left hemisphere) and XI, (right hemisphere). Whilst the observations were in press, I again revised all the brain-specimens and discovered that this fissure was rarely absent.

In most animals (see also Figs. I–III of Recapitulation) this fissure proceeds from the *fossa Sylvii*, or rather from the fissure which we have designated *fissura basilaris lateralis* (blt. in the figs. cited).

In most mammalia the two anterior branches of the *fossa Sylvii* (S″) are absent and therefore our external orbital fissure was mistaken for one of these branches. (For clearness upon this subject we are indebted to an important dissertation by Broca, to which we shall revert.)

Where the two anterior branches of the *fossa Sylvii* (S″) are developed, as in man, the external orbital fissure appears as a third branch, and when it continues to the anterior extremity it separates the orbital lobe (Ob) from the lower and middle frontal gyri (F. 3, F. 2).

The following conditions seem to express a type of the external orbital fissure. Under the lowest border of the M. of the frontal lobe there is a short (3d) incision from the *fossa Sylvii* constituting the posterior portion of the external orbital fissure. This portion is often entirely absent or is only indicated by a shallow indentation.

At the middle part of the external orbital fissure there appears a seldom-failing fissure which is situated somewhat higher than the posterior part and, according to the position of the M. of the frontal lobe, runs more or less horizontally or diagonally (from below and behind in a direction upwards and forwards.)

This middle portion of the external orbital fissure may communicate with the posterior portion. Often, however, it has an independent communication with the *fossa Sylvii*, in which case the posterior portion appears as a fourth anterior branch of the *fossa Sylvii*.

The anterior portion of the external orbital fissure separates the *gyrus frontalis medius* (F. 2) from the orbital fissure. It frequently communicates with the middle portion, and it is comparatively rare that these two portions united communicate, either directly or through the posterior portion, with the *fossa Sylvii*.

The anterior and the middle portion often communicate with the *sulcus frontalis inferior* (f. 2).

In conclusion I would call attention to the noteworthy fact that in Observation V (left hemisphere, p. 38), there is no connection between the *parieto-occipitalis* (po) and the *fissura calcarina*. (cl) (*similar to the ape*).

II.

The observations given and the deductions therefrom have perhaps determined two things with absolute certainty, to wit: 1. THE NECESSITY OF ESTABLISHING A TYPE OF CONFLUENT FISSURES; 2d, THAT THE BRAINS OF CRIMINALS WHICH WE HAVE PRESENTED BELONG TO THIS "SECOND TYPE" (See Introduction, VI, page 54, etc.).

For many of the descriptive details here given, such as are absent in all previous cerebral representations, we are indebted to the special attention which I have bestowed upon these brain-specimens. This keen observation I owe to a casuistic principle heretofore too much neglected by cerebral-anatomists.

In endeavoring to describe any given brain, great numbers of details are observed which are difficult to delineate. In some brains we encounter an exhibition such as in other brains, at least escapes observation.

If now we revise these other brains in this respect, then this exhibition becomes here and there more or less plainly expressed, and it is soon learned that almost every detail has its significance, which in proportion to the strength of development takes a higher grade of form. An example of this is the manner in which the development of the external orbital fissure, in a few brains, lead me to the proof of its existence in all brains, and the great meaning which this fissure has for study of the comparative anatomy of the brain will be shown in a following section. The casuistic mode has given me altogether an incomparably better idea of the mammal brain than came from long years of study.

It may indeed be affirmed that architecturally there exists no fissure arrangement (idea) in the animal brain which has not been expressed in the human.

In the descriptions of our brain-specimens, the deviations from the ordinary method have certainly resulted in but small degrees from increased acuteness of observation by reason of casuistic process.

It hardly need be remarked that the conditions primarily observed upon the brains of criminals need not necessarily belong exclusively to them.

These brains, besides being those of criminals, have the further peculiarity that they belong to the most diverse races and tribes (stems). Amongst them are Magyars, the most distantly related tribes of Sclavonians, Roumanian, one German, and one Gipsy. It might be possible to suppose that the type which we have deducted from these brains is exclusively a fact of comparative race-anatomy, as the normal types of most of these races and clans are unknown, and moreover, there exists no comparative race-anatomy of the brain. But that such is not the case, at least exclusively, seems evident from the fact that such an eminent judge of the brain as Betz, of Kieff, who has naturally made his studies chiefly upon Sclavonic brains, has acknowledged the atypy of my specimens, and in this manner he has performed an active part in the accomplishment of this work. Zernoff's[1] work proves that many of the conditions may be more frequently found in the brains of the Sclavonians,—though not so frequently as in our brain-specimens,—than in those of the Germanic and Latin races.

In these brains (of criminals) which were perhaps conspicuously Sclavonic, it seemed especially noticeable that the *sulcus interparietalis* communicated very frequently with the *fossa Sylvii*, and the *parieto-occipitalis* (po) with the *horizontalis* (ho) and the *interparietalis* (ip).

[1] Zernoff's (of Moscow) work upon "*The Individual Types of the Convolutions of Human Brains,*" appeared in the Russian language. No book is quite sealed to me with seven seals though it may be with some. Therefore, I must beg to be excused if I overlooked or wrongly read some points.

Zernoff "seldom" observed a union of the frontal sulci with the *fossa Sylvii*. In his collection of one hundred brains, the *sulcus centralis* seems to have had connection in but one instance and then with the *fissura Sylvii*.

He seems to have observed a connection with the *scissura hippocampi* (h) with the *calloso-marginalis* (cm), the *parieto-occipitalis* (po), and *collateralis* (cl) and of this last with the *occipitalis* (po); the same also as regards a communication between the *calloso-marginalis* (cm) and the *parieto-occipitalis* (po).

At all events, I hope that these very questions which I here present for debate, will serve as a spur to somewhat promote a knowledge of brain anatomy as it exists in European races; and it especially devolves upon the Austrian, Hungarian, and Russian physicians, to complete this work.

The authority of Betz and the work of Zernoff enable us to reject—as being insufficient at least—the idea that the specimens of criminals' brains which we have portrayed, represents but a deviating type of a normal Sclavonian brain.

The fact also that in the brains of five different races such great deviations from the normal type are found in common, forbids us, a priori, to consider these brains as expressing no more than deviations incident to comparative race-anatomy.

There remains nothing more, for the present at least, but to express the proposition:

THE BRAINS OF CRIMINALS EXHIBIT A DEVIATION FROM THE NORMAL TYPE, AND CRIMINALS ARE TO BE VIEWED AS AN ANTHROPOLOGICAL VARIETY OF THEIR SPECIES, AT LEAST AMONGST THE CULTURED RACES.

This proposition is calculated to create a veritable revolution in Ethics, psychology, jurisprudence, and criminalistics. For this very reason it should be handled with the greatest prudence; it should not yet serve as a premise; and for the present it should not leave the hands of expert anatomists.

In matters of fact it must yet be repeatedly proven and that from many different points of view, until it can finally rank as an undoubted addition to human science.

The variety of conditions which we may expect to meet in

the different races, will assign to this proposition a little halting place in the history of Science, and worthless as well as valuable contributions will, for a time to come, give rise to oscillations of opinion.

It is self-evident that the observations here collected are the result of an *a priori* conviction that the constitutional ("eigentliche") criminal is a burdened ("belastetes") individual; that he has the same relation to crime as his next blood kin, the epileptic, and his cousin the idiot, have to their encephalopathic condition (and its results. Tr.).

Even in these this encumbrance does not signify actual disease, but a predisposition to it only.

As I desire no conclusion of this question until it is anatomically solved, I will not return to the grounds of a natural psychology which led to an *a priori* proposition. I shall equally abstain from offering to explain the facts of empirical criminal psychology and also those statistics which are calculated to support this proposition.

In this direction I would, along with other citations, refer to two of my previous essays: 1st, "PSYCHO-PHYSICS OF MORALITY AND RIGHT," Vienna, 1875. Pub. by Urban & Schwarzenberg.

2d. "NATURAL HISTORY OF CRIME," *Wiener juristische Blätter*, No. 1 to 3, 1876.

These studies elicited a violent opposition. I do not mean the local one; that can be understood and judged of in a great degree, not from the intellectual standpoint but exclusively from an ethonopathic view.

The various grounds of opposition afforded me much material for thought, and I will ask the question, was the violence of this opposition in any way justified? I have no desire to recriminate, I only wish to assist towards a clear understanding.

Since Kant (1781) established his doctrine of antinomies the honest friends of truth know that there exists a quantity of metaphysical theses and counter-theses which, no matter how contradictory, human reason must for the present look upon as equally defensible. To these propositions belong the doctrine of Freewill and that of Predestination (3. Antinomie).

We may have the conviction that freedom of mind is but the expression of wealth of mind, that all mental actions are expressions of fixed natural laws. In the same way also we may acknowledge an absolute psychical freedom. But the antinomies of Kant are the goal of knowledge, they cannot be its premises.

In regard to antinomies, therefore, the *a priori* attitude of science is that of metaphysical neutrality.

By his Criticism of Pure Reason, Kant freed humanity from the scourge of metaphysical intolerance; yet, after a lapse of almost a century, humanity has not entered upon this inheritance. The past century has handed down to us the principle of religious toleration, which is already infused into the flesh and blood of all civilized states.

I do not think to deceive myself when I take for granted that our century will soon succeed also to that other inheritance; metaphysical neutrality.

But first, scholars must place themselves decidedly upon this ground, then will follow society, after that governments and finally legislation.

There will still exist a period of renunciation, as on the one side of a line of antinomies (concealed power), there will be found position, decorations and titles, and on the other side either open or secret social persecution; and neutrality will be regarded as opposition.

III.

In examining human brains that we have reason to look upon as of low grade, we at once discover "animal similarities." To the present point we have confined ourselves chiefly to proving the similarities with the ape-brain ("ape-similarities"). But a close comparative study of the human and mammalian brain led to the query whether there was an essential difference in the construction of the mammal brain as opposed to the human and ape brains. In two previous communications[1] I have put this question as one to be negatived, and I refer to it more especially as nature has expressed hardly an idea in the mammal brain which is not, under certain circumstances, repeated in man. It can be comprehended that it is quite impossible to speak of a psychological similarity to beast of prey, apes, &c., when an architectural characteristic occurs in a human brain which is found to be typical in those animals.

I have already given special emphasis (loc. cit.) to the fact that the difference in the structure of the frontal human brain, with its three primitive gyri, and the frontal brain of certain beasts of prey with four primitive gyri (e. g. the fox), is only a seeming one.

In man the upper two frontal gyri blend together, and their separation is indicated only by the secondary sulcus φ. (Fig. I of Introduction.) In some human brains (see for example Maglenov, left, Obs. VI,) there is developed from this

---

[1] See "*Types of Beasts of Prey in the Human Brain.*" Centralblatt für die medic. Wissensch. 1875, No. 52, and "*The Occipital Lobes of Mammalia,*" in the same, 1876, No. 10.

secondary fissure a long and deep fissure which extends parallel with the superior frontal sulcus, and then the type of four primitive gyri is exhibited in the human brain.¹

Fig. I represents (enlarged) a fox's brain. The lower three primitive gyri in the frontal part are designated as (F. 3, F. 2, F. 1) and the dividing fissures as (L 2, L 1). The upper one of the three parallel fissures is designated by φ and the gyrus which lies above it with ♦. A fissure (a) which unites with the *fossa Sylvii*, separates the frontal from the orbital lobe (Ob) which here comes to lie upon the outer surface. This fissure corresponds to the external orbital fissure in man, and in apes this is more developed than in man.

As before mentioned (p. 154, &c.) it is exceptional that this fissure is well developed in man, and it is found as an extension of a third and lower branch from the *fossa Sylvii*, which penetrates into the foremost end of the anterior brain. We have (loc. cit.) also before directed particular attention to the fact that the fissure (b) of the fox's brain represented the *fissura olfactoria* (of) in man. It also happens in man that the middle gyrus (F. 2) and the lower part of the upper frontal gyrus (F. 1) blend, the upper gyrus being divided by a deep fissure into two parts. In this case the upper frontal sulcus (L 1) is undeveloped and φ takes its place. (See Obs. XIII.) We have here in the human, an instance of the blending of the two middle gyri (F. 2 and F. 3) of the typical four primitive gyri of beasts of prey, whilst the upper gyrus is, as it were, free.

That which especially distinguishes the frontal brain of primates from the lower mammals is the absence in the latter of radiating fissures.

In Fig. I, of Recapitulation (Fox), at first glance, at least, all radiating fissures of the frontal brain are absent, and more especially, 1st, The radiating branch of the inferior frontal

---

¹ M. HANOT found in four instances out of eleven cerebral autopsies of *Vieux vividiviers* ("*veritable piliers de prison*") four frontal convolutions instead of three. In these four cases the middle frontal convolution (F. 2) was the one doubled.

M. OVION did not find a single like-instance during the year 1879, among those who died at the Hospital Cochin, and they are rare in hospitals generally. (Tr.).

sulcus (f. 2) which is the third frontal sulcus (f. 3) (perpendicular). 2d. The radiating fissure of the superior frontal sulcus (f. 1). 3d, That of the fissure φ. 4th, The *sulcus centralis* (c). We will first enquire how these radial fissures arise.

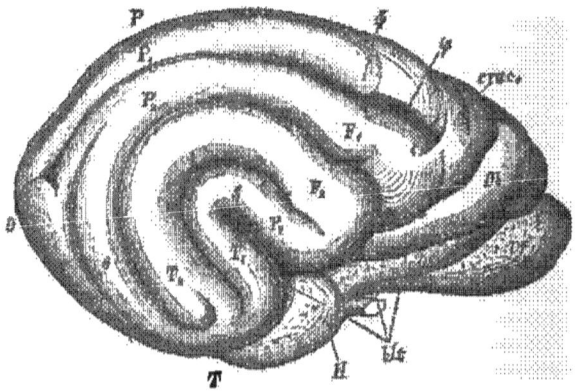

FIG. L.

Fox (*enlarged*).

F. 1, F. 2, F. 3.=1st, 2d, and 3d frontal gyri.
P. 1, P. 2.=1st and 2d parietal gyri.
T. 1, T. 2.=1st and 2d temporal gyri.
S.=Fossa Sylvii.
Ob.=Orbital lobe.
Ol, Of.=Olfactory lobe.
O, O.=Occipital lobe.
P.=Parietal lobe.
T.=Temporal lobe.
Cu.=Cuneus.
H.=Gyrus hippocampi.
φ.=Upper (4th) frontal gyrus.
f. 1, f. 2.=1st and 2d frontal sulci.
a.=External orbital fissure.
b.=Sulcus olfactorius.
c.=Sulcus centralis.
φ.=Upper (secondary) frontal sulcus.
blt.=Fissura vasilaries lateralis.

If we study our brains of criminals it conducts us first of all to the knowledge of an important law. That is, where a sagittal fissure is interrupted, a radial fissure is formed, so that, for example, when the upper frontal sulcus is divided into several parts, at the posterior end of each piece a radiating fissure is formed (See especially Obs. XIV).

These radial fissures may become separated from their sagittal stem, as is often to be seen in case of the third frontal sulcus (f. 3) and its sagittal stem, the lower frontal sulcus (f. 2).

Here enters the question, whether the *sulcus centralis* has not originated in this manner, and by further development become a separated, radial branch of one or all of the frontal sagittal sulci?

The facts observed in one criminal's brain are strongly affirmative, as we there find connections between it and all the sagittal frontal fissures (φ, f. 1, f. 2).

Figs. I and III (Recapitulation), brains of Fox and Bear, actually do show such radiating fissures (c). In the Fox's brain we observe that it (c) bounds anteriorly the fissure of the upper primitive gyrus, and in the Bear it extends from the anterior part of the fissure of the upper primitive gyrus in a direction backwards and upwards.

We must now search for further proofs as to whether the radial fissure marked (c) in these two cuts really correspond to the *sulcus centralis* (fissure of Rolando).

We should first observe that nearly all gyrocephalic mammalia exhibit a distinct paracentral lobule.

In many animals the anteriorly convex curve which the *sulcus calloso-marginalis* sends off to the external surface in its course from behind forwards, forms the *fissura cruciata* of Leuret (See Fig. I, cruc.), and this *fissura cruciata* forms also the anterior curve of the arch-shaped paracentral fissure.

This is the case for example in the Fox.

In other animals the *fissura cruciata* proceeds from the separated frontal portion of the *sulcus calloso-marginalis*, and the anterior-bordering curve of the paracentral lobule connects with this *fissura cruciata*.

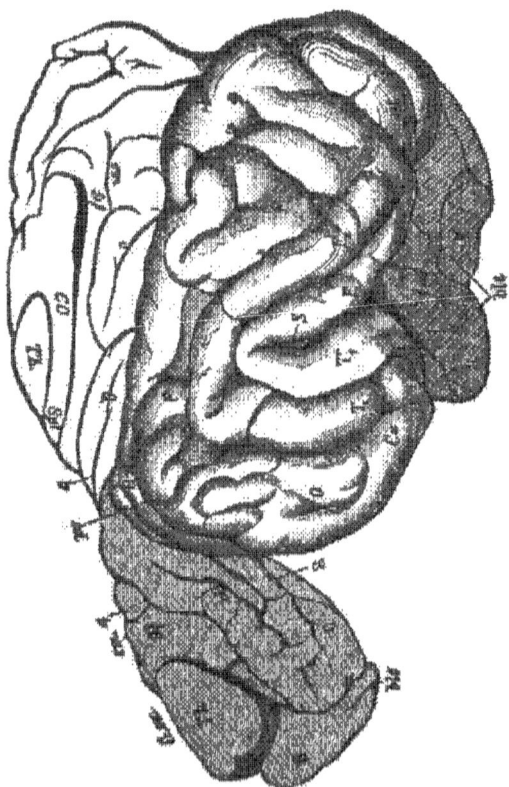

FIG. III.
BEAR.
(With the medial and basilar surfaces turned and brought to view.)
F. 1, F. 2, F. 3.—1st, 2d, and 3d frontal gyri.
O. O.=Occipital lobe.
P.=Parietal lobe.

T. 1, T. 2.= 1st and 2d temporal gyri.
S.= Fossa Sylvii.
Ob.= Orbital lobe.
Ol.= Olfactory lobe.
Cu.= Cuneus (and G. cu).
H. H.= Gyrus hippocampi.
Q. Q.= Praecuneus.
Gf. Gf.= Gyrus fornicatus.
CC.= Corpus callosum.
Th.= Thalamus opticus.
Spl.= Splenium.
♦.= Upper or 4th frontal gyrus.
a.= External orbital fissure.
b.= Sulcus olfactorius.
c.= Sulcus centralis.
f. 1, f. 2.= Upper and lower frontal sulci.
♀.= Upper frontal (in human, secondary) sulcus.
cm. cm.= Sulcus calloso-marginalis.
cl.= Sulcus collateralis.
po.= Fissura parieto-occipitalis.
q. q.= Fissure between basilar lobe and gyrus fornicatus, and which connects with the sulcus calloso-marginalis.
blt. blt.= Fissura basilaris lateralis.

Again there are animals (the Bear for example) in which the anterior separated branch of the *sulcus calloso-marginalis* also furnishes the *fissura cruciata* (l., in Fig. III), but in this case the anterior curve of the fissure which limits the paracentral lobule is placed back of the *fissura cruciata* and is not connected with it.

However this may be, a radiating fissure, corresponding to the *sulcus centralis*, must be situated back of the *fissura cruciata* of Leuret.

Both of the fissures (c) in Figs. I and III correspond to this.

Where this radiating fissure extends into the paracentral lobule, it must in its course separate the lobule into two equal parts.

Fissure c in Fig. III represents this condition.

No matter how common the existence of a double curve of the *sulcus calloso-marginalis*, one of which is convex anteriorly and the other posteriorly, and from which would result an appearance that would correspond entirely with the paracentral lobule, it might still be urged that possibly it signified a similarity but not an identity.

Identity can be proven, in the first place by histology. If the fissura (c) of Fig. I and III (Recapitulation) really correspond to the *sulcus centralis*, then the main deposits of Betz's giant cells, and the heaped nests of the same, must lie in front of the fissure (c), and the gyri before and back of the limiting fissure of the paracentral lobule must correspond to the anterior part of the paracentral lobule, and to the highest part of the *gyrus centralis anterior*.

Furthermore, Betz's pyramidal giant cells ought also to disappear at the upper part of a transverse section made at the posterior limit of the paracentral lobule. Investigations, so far as concerns the Fox and Bear, verify these suppositions.

A transverse section through the *fissura cruciata* of the Fox brought to view masses of Betz's giant cells in the upper gyrus.

Transverse sections through the middle of the paracentral lobule of the Fox and through the anterior point of the anteriorly convex curve of the fissure (c), (Fig. I), and sections made between these and the first, exhibited masses of giant-cells in all the frontal gyri quite down to the orbital lobe (Ob).

In a transverse section made just back of the posterior edge of the paracentral lobule, no giant-cells were anywhere found.

By these means the existence and position of the *sulcus centralis* (c) and of the paracentral lobule in the Fox is strictly proven.

Similar transverse sections were also made in the Bear's brain. The first, through the *sulcus cruciatus*, c of Fig. III (Recapitulation). This section passed through fissures (a) and (b), and giant cells were found in the frontal gyri.

The section passed through the bend of the upper fissure (v) and the fissure (c), and on its entire upper and external border, down to the orbital lobe there were great quantities of pyramidal giant cells.

The third transverse section passed through the posterior portion of the paracentral lobule which appears very well defined in Fig. III. In this section only isolated groups of pyramidal giant cells were to be found, in the vicinity of the posterior curve of the paracentral fissure.

This then, proves the right, in case of the Bear, to look upon the elliptical gyrus into which fissure (c) (of Fig. III) penetrates, as the paracentral lobule, and the fissure (c) as the *sulcus centralis*.[1]

By the way, this also teaches how to define the frontal lobe.

If, in Fig III, an imaginary line be drawn from the point where the fissure (c) bends upon the inner surface to the *fossa Sylvii* (just under the place in the Figure where the letter F. 3 starts) it will indicate the posterior boundary of the frontal lobe.

A like procedure must be followed in case of the Fox (Fig. I) and this would show that the marks F. 1, F. 2, F. 3, f. 1, L 2, in the cut are placed too far back.

We have before spoken about the importance of fissures (a) and (b) in Figs. I and III and of their representatives in the human brain (See Recapitulation, p. 140, &c.).

In the frontal lobe of the mammalia then, there are absent only the two frontal branches of the *fissura Sylvii*, which appear first in the ape (There is such a fissure in the horse (see Fig. II between (a) and (S) and the radial branches of the sagittal frontal sulci (L 2, f. 1, and φ).

As important as the first, especially, may be, it still would not lead any one to deduce from this condition a qualitative difference between the brains of primates and other mammalia.

The question is entirely undecided whether the gyri an-

---

[1] The celebrated work of Betz ("ANATOMICAL PROOF OF TWO BRAIN CENTRES," *Centralblatt*, 1874. Nos. 37, 38) had already really decided the question in our favor as to the supposed position of a *sulcus centralis*, and it is also decided in the same manner through a more recent treatise by Bevan Lewis ("BRAIN," Journal of Neurology, part I) on "THE COMPARATIVE STRUCTURE OF THE CORTEX CEREBRI."

terior to the *fossa Sylvii* and posterior to the just-mentioned line limiting the frontal lobe, marked F. 2, F. 3, are to be considered as belonging to the temporal or parietal lobe.

That which has usually been assigned to the temporal and to the parietal lobes through topographical considerations, is indicated by the letters, P. 1, P. 2, and T. 1, T. 2.

Let us now examine the base. In the anterior basilar lobe of mammalia it is particularly observable that the olfactory lobe is much more developed than in primates and especially than in man; there the *fissura Sylvii* does not extend to the base but is represented by a much less marked "deepening" (the real *fossa Sylvii*) and it is also to be seen that the orbital lobe is pushed to the external surface. The *fissura olfactoria* (b) in Figs. I, II, and III, is generally connected with the *fissura Sylvii* or another fissure (*fissura basilaris lateralis*, blt. in the Fig.) with which we shall soon become acquainted. The *fissura orbitalis* (ob) is generally absent, though it is represented in the horse (Fig. II). The middle basilar lobe is generally undivided. (In the horse, however, for example, the division is indicated). See Fig. II. The same as in man the basilar lobe, in the middle cranial fossa (*gyrus uncinatus* (U) and *gyrus Hippocampi*, (H) middle basilar lobe) gives off a lobe which rests in the posterior cranial fossa as the occipital basilar lobe. This is the case with all mammalia (Fig. II). In man this basilar occipital lobe is divided into two parts by the *sulcus collateralis* (cl) as this massive development demands a highway for vascularization.

In many animals this sulcus (cl) is absent, in others, as in the Cat and wild Boar, it is more or less plainly present (See for example, cl. in Fig. III, of the Bear's brain). This shows that the mammal also has its basilar occipital lobe and which in several species is even separated into the *gyri lingualis* and *fusiformis*. This portion of the mammal brain I have not taken at all into consideration (s. l. c.). Before I commence with the external surface of the occipital lobe of mammalia, I must consider a fissure which appears in most gyrocephalic mammalia and which I prefer to designate as the *fissura*

RECAPITULATION III.

basilaris lateralis (blt.) (see Figs. I-III, Recapitulation). Fig. II exhibits this fissure in the Horse. It is seen to be connected with the *fossa Sylvii* (S). It sends, 1st, anteriorly the branches (a) and (b); 2d, posteriorly, it first separates the external temporal surface (T) from the middle basilar lobe. This part of the *fissura basilaris lateralis* is indicated in Fig. II with (fs). The fissure then extends further back

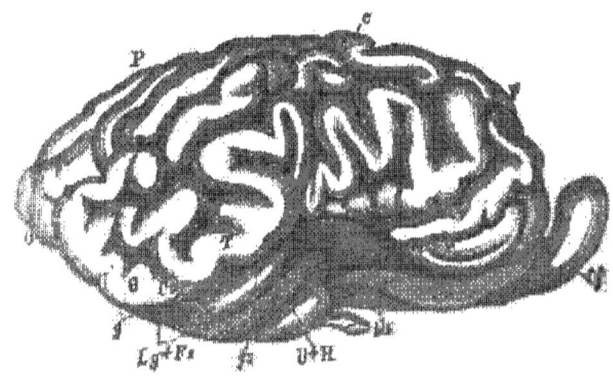

FIG. II.
HORSE.

F. = Frontal lobe.
OOO. = Occipital lobe.
P P. = Parietal lobe.
T. = Temporal lobe.
S. = Fossa Sylvii.
Ob. = Orbital lobe.
Ol. = Olfactory bulb.
Cu. = Cuneus.
Lg. + Fs. = Gyri lingualis and fusiformis.
fs. = Sulcus fusiformis.
blt. = Fissura basilaris lateralis.
U + H. = Gyri uncinatus and hippocampi.
a. = External orbital fissure.
b. = Sulcus olfactoria.
c. = Sulcus centralis.
d. = Sulcus orbitalis.
g. = Sulcus oc.

in order; 3d, to separate that part which I regard as the external surface of the occiput (O) from the occipital basilar lobe. In Fig. II this part is indicated by (g). In many animals the posterior portion of this fissure is bifurcated. (See Fig. III, fissure (cc).)

This fissure also separates more or less completely: 1st, with its occipital extremity, the external surface from the occipital basilar lobe; 2d, with its middle portion, the external surface from the middle basilar lobe, and 3d, the frontal lobe from the orbital.

This fissure, in some sections torn apart, is represented in man; the first-mentioned portion (posterior) is represented by *Wernicke's fissura occipitalis inferior* (g) (see Fig. I, Introduction); the second (middle) by *Wernicke's fissura fusiformis*, which generally corresponds to our *fissura temporalis tertius* (ibidem) and the anterior part by the external orbital fissure in man.

In the brains of criminals which we have presented, the *fissura basilaris lateralis* is very beautifully demonstrated.

Through the union of the *fissura occipitalis inferior* and the *fissura fusiformis* and the full development of both, they often furnish a good representation of the first and second portions (posterior and middle) of the *fissura basilaris lateralis* (see for example, Obs. XV).

Our specimens teach, moreover, that the upper frontal branch of this *fissura basilaris lateralis* sometimes appears complete in man (See for example Obs. IV, p. 34; Obs. X, p. 55; Obs. XI).

It remains to be mentioned that the *fissura basilaris lateralis* does not necessarily join with the *fossa Sylvii*; they are often separated by intervening gyri. I have spoken of this in a previous communication.

We will now turn to the external surface of the occipital lobe.

We can at least say that the portion which pertains to the basilar occipital lobe, that is, the posterior portion of the under surface and the posterior edge of the external surface and its contiguous parts, belong to the occipital lobe.

In primates we observe moreover the parietal lobe (P) directed posteriorly upon the substratum of the temporal lobe (T), located partially around the posterior extremity and ending on the inner surface at the *præcuneus* (Q), above the occipital basilar lobe.

In other mammalia the brain is more elevated at the posterior extremity; the parietal lobe bends down under.

We have then every reason to look upon all that which is situated behind that portion of the parietal lobe which bends around the temporal lobe (T) as the occiput.

The lowest point of this curved convolution I have termed *Cuneus* (Cu) (see Fig. I and III, Recapitulation).

This, however, will scarcely be verified.

If we speak of an occiput there must also be internal evidences of it; and to these belong a central ganglion connecting with the posterior part of the *cornu ammonis*, corresponding to the *cornu ammonis posterior*, and to the *calcar avis* in primates.

This I have found in all mammals of the most varied classes. Up to the ape it has a more or less oval form, is in bulb-like, concentric layers, composed of white and gray substance, and is undoubtedly a representative of the *cornu ammonis posterior* in primates.

By this, the existence of an occiput, and that a very considerable one, too, is proven.

The existence of a central vascular canal of the occiput—the posterior cornu—I have found in most of the animals examined. It is sometimes absent, as for example, in the Bear.

There it is represented by deep fissures entering from the outside, for example (cc) and (a) in the Bear (see Fig. III, Recapitulation).[1]

---

[1] I have already mentioned, in another place, ("ANATOMICAL INTRODUCTION TO BRAIN DISEASES," in my Neuropathology and electropathy), that the relative preponderance of the *cornu ammonis* over the other central ganglia, and especially over the *corpus striatum*, is a prominent characteristic of the mammal as contrasted with that of man; and this statement was verified by MIHALKOWIC in his excellent work upon the "THE DEVELOPMENT OF THE BRAIN."

The fissure indicated by (g) in Fig. II (Recapitulation I) have spoken of as corresponding, in the human brain, to *Wernicke's fissura occipitalis inferior*; and that indicated by (cc) Fig. III, as corresponding to the *fissura calcarina*.

It is certain that the lowest part of this fissure corresponds to the inferior occipital fissure (g). In many animals (the Fox, for example) this fissure takes the place of the *fissura calcarina* in man.

This is the fissure through which the posterior cornu is vascularized and in sections made successively from behind towards the front, it diminishes in ratio to the greater development of the posterior *cornu ammonis*.

A glance at Fig. III (Recapitulation) shows also that when the occiput turns inwards and rests on the occipital basilar lobe, and anteriorly against the *praecuneus*, as is the case with primates, the upper part of this fissure is then torn away from the lower, and to a certain degree assumes the office of the *fissura calcarina*, that is, so far as being the separating fissure between the *Cuneus* and the occipital basilar lobe, and with its upper portion, which I have marked (po), it divides the *praecuneus* (Q) from the *Cuneus* as does the *fissura occipitalis perpendicularis* in man.

Another instructive fissure is the one which I have designated (q) (see Fig. III, Recapitulation).

It forms, as does the common stem of the perpendicular occipital fissure and the calcarina in man, the dividing fissure between the occipital basilar lobe and the *gyrus fornicatus*, and communicates with the *sulcus calloso-marginalis*.

We now comprehend the great importance of that condition in our brains of criminals, where the *sulcus calloso-marginalis* extends through the *praecuneus* and unites with the stem of the *fissura occipitalis* (po).

This fissure (q) also sends an extension to the *praecuneus* (Q) as we have so often seen the *parieto-occipitalis* do in the specimens of criminals' brains, by which it may become united with the fissures of the *praecuneus*, or the branch may remain independent of them.

In the Bear's brain the fissure (q) and the upper part

of the fissures cc and po, fulfil the same purpose that the united Y-formed calcarina and parieto-occipital do in man, and which two last, in the ape, are not yet united.

On the other hand, we see that already in the Bear one of these fissures connects with the lower occipital fissure; but in the bear these fissures (cc) and (q) form a *nidus* with the *collateralis* (cl) and this again makes plain the significance in the criminal-brain of the frequent union of the lower occipital, the *collateralis*, the *calcarina* and the *parieto-occipitalis*, with each other, and of the latter with the *calloso-marginalis* (cm).

Thus the study of our specimens makes the facts of comparative anatomy distinct and these in return demonstrate the atypic character of our brain specimens which, by deduction, may be justly termed retrograded brains.'

From these facts results the following important proposition:

There exists no qualitative difference between the brains of mammalia and those of primates.

NOTE.

This proposition is the result of a comparison of numerous transverse sections which I have years ago made of the brains of the most widely differing animals, and of the most varied classes of mammalia; and they have served me as a guide in studies of the external cerebral surface. I have a large collection of transverse sections of the occiput, from the maggot to the common bat. In them it can be seen that mammals without convolutions are not constructed on a different plan from the gyro-encephalic mammalia. The fissures have only a haemato-dynamic significance. That is, they prevent vessels with considerable arterial force from entering directly into the mass of the brain. Where the quantity which is to be received from the surface is small, i. e., not penetrating deeply, no fissures are required.

Where, in the ascending animal scale, any part is more strongly developed, a new typical fissure appears.

We will here once more emphasize the fact that a frequent connection of a typical fissure signifies no enriching of the fissure, but simply the absence of developed gyri.

It may also be emphatically stated that fissures do not merely signify a separation; they are more significant of a focus for the entrance of nourishment, vascularization, a consideration which is of special importance in cerebral pathology.

' Corresponding to the fact that the fissure through which the posterior *cornu* *ammonis* receives nourishment (is vascularized), is situated in many animals (the Fox for example) outside instead of inside, the occipital ganglion itself submits to an axial deviation outwardly, and the posterior cornu, when it exists, to an inward deviation; as I have already mentioned in a previous communication (l. c.).

I do not consider the representations which we have here given as containing all the proof in support of the proposition; on the contrary, I am preparing to resume the subject, accompanied by co-workers, in order to make exhaustive researches and to investigate especially the differential histological characteristics of individual convolutions.

Whilst this section was prepared for publication, I became acquainted with two important works by the celebrated French Naturalist, Paul Broca ("*Anatomie comparée des circonvolutions cérébrales. Le grand lobe limbique et la scissure limbique dans la série des mammifères.*" Revue d'anthropologie, 1878; and "*Etude sur le cerveau du Gorille,*" ibid.).

The first work, more especially, is a direct continuation of the labors of Leuret and Gratiolet, and is equal to them in value. By its clearness of presentation, its resultant tracing of an important anatomical truth, and its multiplicity of material, it will occupy a place of first importance in the science of cerebral anatomy.

Broca's results frequently contradict my representations. Notwithstanding this, I do not think it best to remodel this section of my work, but rather to communicate the results of my studies, as they appear to me, because many important statements of Broca have not sufficiently impressed me, and it is my opinion too that the greater the contrast in the results of different authors, derived from the same material, the easier in the midst of these variances will truth be discovered.

I hope that we will be assisted to this ultimatum by Betz, who will shortly enter the discussion with a larger work.

Broca's starting point is the "*lobus limbicus*" (encircling lobe). This term is applied to that formation found in many animals (the Otter for example), which is a curved lobe made up from the combined *lobus olfactorius*, the middle basilar lobe of animals (*gyrus hippocampi*) and the *gyrus fornicatus*. The external olfactory root connects with the *gyrus hippocampi* and the internal root with the *gyrus fornicatus* and in this manner the upper and lower portions of the curve meet each other anteriorly. This *lobus limbicus*, as a starting point in comparative anatomy, is a real master stroke.

Of equal importance with the *lobus limbicus* is the fissure-system which separates it from the superimposed hemispheric mass ("*Masse circonvolutionnrie*"). By the way, the *scissura limbica inferior* of Broca, runs along the external border of the hemisphere and corresponds entirely to our *fissura basilaris lateralis*.[1]

The upper part of the *scissura limbica* represents the *sulcus calloso-marginalis*, but with this exception: that in primates this fissure generally separates only the anterior portion of the the *gyrus fornicatus* from the remainder of the hemisphere, whilst in other mammalia the curve is more or less complete, if not indeed entirely uninterrupted. In many mammalia (beasts of prey for example) that portion of the fissure which separates the parietal lobe from the parietal part of the *gyrus fornicatus* is further developed and extends with an anteriorly convex curve to the external surface where it constitutes the *sulcus cruciatus* of Leuret. Concerning the relation of the *calloso-marginalis* to the *scissura cruciata of Leuret*, see previous statements.

In our specimens of criminals' brains, Broca's *scissura limbica* is frequently seen to be complete, in so far as it extends without interruption from the frontal part of the *gyrus fornicatus* to its posterior curve, or even communicates with the stem of the united *parieto-occipitalis* and *fissura calcarina*, and that, which is very important, sends upwards two curves which serve as borders for the *paracentral* lobe (Broca erroneously credits Meynert, instead of Betz, with the discovery of this lobe).

If it is preferable, especially on account of the condition found in man, to denominate the lower part of the *scissura limbica* as the *fissura basilaris lateralis*, then the upper part would be better designated as the *fissura calloso-marginalis*.

At the same time, the great value of Broca's expression

---

[1] I prefer my designation because it is more generally applicable. We shall see that there is a great difference of opinion between Broca and myself as regards the representation of this fissure in primates, and which seems chiefly to arise from the reason that Broca was unacquainted with Wernicke's important work, "*Convolution-system of the human brain*" (Arch. für Psychiat, 1876, vol. vi.)

consists in its description of the fundamental condition, and for that reason it ought to be retained.

The upper and lower curve of Broca's *scissura limbica* are posteriorly always separated from each other by a portion of gyrus which extends from the *gyrus hippocampi* to the posterior edge of the external hemispheric surface. Broca calls this bit of gyrus, "*plis de passage rétrolimbique.*" It corresponds to that which I call simply basilar occipital lobe.

Broca, as well as myself, found a fissure in this piece of gyrus which he considers as the fore runner of the *fissura calcarina*, while I take it as the analogue of the *sulcus collateralis*, which Broca does not take into consideration; which is the correct view, the histology of surrounding cortical substance can alone decide.

The upper curve of the *scissura limbica* is generally interrupted in its anterior part, and thereby an upper communication is established between the *gyrus fornicatus* and the *lobus frontalis*. In the horse the upper curve is entirely complete and there exists only a slight interruption between the upper and lower curve of the *scissura limbica*, through which the most anterior and lowest part of the *gyrus fornicatus* communicates with the *lobus frontalis*.

The labors of Broca have imparted the highest interest to a branch of our *fissura basilaris lateralis* (blt) (or the under curve of the *fissura limbica*), which we have termed external orbital fissure (see (a) in Figs. I–III, Recapitulation) and which, by a stretch of gyrus, is often superficially separated from the *fissura basilaris lateralis*.

According to Broca, this branch corresponds in mammalia to the *sulcus centralis* (fissure of Rolando).

This view of the celebrated French anatomist possesses a certain fascination; nevertheless, it is decidedly incorrect.

Broca's *sulcus centralis* lies before and below the center of Betz's pyramidal giant cells, whereas the veritable *sulcus centralis* must lie behind and somewhat in the midst of those centers.

There can be no doubt that this fissure represents an external orbital fissure.[1]

I most decidedly agree with Broca that the anterior branches of the *fissura basilaris lateralis* have nothing to do with the anterior branches of the *fossa Sylvii* in man, and he justly characterizes the statements of Meynert as "*confusion.*"

Before I consider what becomes of the *fissura limbica* in primates, I will make an observation on the *lobus temporalis* and the middle basilar lobe.

Broca really allows in the mammalia (excluding the primates), only the lowest piece of the posterior part of the second super-sylvian convolution-arch as an indication of the temporal lobe as found in primates, whereas I, from purely external topographical reasons, claim for it, in mammalia, the entire lower posterior part of the lowest two super-sylvian convolution-arches.

Present methods cannot decide which view is correct.

Broca's idea respecting the *gyrus uncinatus* is much more extraordinary, a view which he treats as self-evident, and so does not even distinctly formulate it.

The French anatomist does not regard the middle basilar lobe in mammalia as analogous to the middle basilar lobe in primates; he rather holds the *gyrus hippocampi* in man as the representative of the middle basilar lobe in non-primates, and the observation made by me that the *gyrus uncinatus* frequently represents only the turning loops of the temporal gyri, and the previous observation by Wernicke that the development of the *gyrus uncinatus* depends entirely upon the development of the temporal gyri, favor the French

---

[1] I will here direct attention to an error which can easily find entrance in connection with the Sheep's brain and which has indeed found way into literature. I have before me an interesting Italian paper containing an account of the experiments at Siena in the physiological laboratory of Prof. Albertoni (Milan, published by Rechiedei, 1876). Albertoni communicates his important experiment upon non-ætherized dogs and cats, for a real epileptic center at the upper border of the *fissura postruciata*. This little book contains also a treatise by Student Marcacci, upon the motor centers in sheep. He placed the centers all in front of the *fissura cruciata*. But he overlooked the real *fissura cruciata*, which is shallow and rises from the *sulcus limbicus*, and mistakes for it the deeper, well characterized *sulcus centralis*.

naturalist's idea. Broca observed a fissure in the lower human races which separated the *gyrus hippocampi* from the *gyrus uncinatus*; he assigned it to the temporal lobe and bestowed upon it no special name; this fissure he takes as the representative of the middle part of his *scissura limbica inferior*.

Our specimens have shown this fissure dividing the two gyri of our middle lobe to be very often present, but it is to be seen that it chiefly appears as an extension of the *sulcus collateralis* (cl) (see for example, Fig. III, Plate XII).

I, on the contrary, regard *Wernicke's fissura fusiformis* (t. 3) which separates the *gyrus uncinatus* and in part, the *gyrus fusiformis*, from the external surface of the temporal lobe, as a remnant of the middle portion of the *fissura lateralis basilaris*, or *fissura limbica inferior*. In order to determine which view is correct, deciding data must be produced.[1]

Broca belows no attention upon the posterior part of the *fissura basilaris lateralis*. I have already mentioned that I consider *Wernicke's fissura occipitalis inferior* (g) as the remnant of it, and the fact of the frequent blending of this with the *fissura fusiformis* is surely of importance.

Again I specially repeat that I formulate these counter propositions so plainly, because I believe that the more distinctly variations of ideas are opposed to each other the more promptly truth will gain footing.

Still more diverse are our views respecting the *lobus occipitalis*.

Broca says that it is in those animals where the temporal lobe exists that the earliest representation of an occipital lobe is to be found (corresponding, by the way, to Cn of our Figs. I to III, Recapitulation.) But further on in his work he devotes no more attention to this statement made in the early part. Through a folding of the parietal lobe, from whence results the *fissura parieto-occipitalis*—the occipital lobe, in primates,—should be separated from the parietal lobe, the

---

[1] The tendency which *Wernicke's fissura fusiformis* has to unite with the extension of the *collateralis*, which separates the *gyri uncinatus* and *hippocampi*, is certainly noteworthy. This fact is testimony in favor of Broca's idea.

occipital lobe comprising that part which lies behind the fissure. Where this fissure is absent there is, according to Broca, also an absence of the posterior cerebral "pole." I deem it more correct to claim for those animals in whose brains this fissure is absent, such parts as correspond topographically to the occipital lobe in primates; —that is, the bent-under portion of the upper part of the parietal lobe at the posterior "pole". Illustrated by means of sections, I have, in a previous contribution, given my reason for the view I entertain, namely: the existence of an occipital central ganglion, and when the eminent French anatomist, to the solution of this point, has devoted a few brains to the knife, I have no doubt that he will at once be converted to my view (see Figs. I–III of Recapitulation).

It is highly important to study the extent which Broca assigns to the parietal lobe. He so denominates the entire area of external hemisphere which lies above and back of the *sulcus centralis* (a) (in our Figs. I–III).

He allows only a trace of the temporal and occipital lobes above the *fissura basilaris lateralis* (blt. ibidem), whereas, our temporal lobe extends quite to this fissure.

I have already called attention to the fact that a large piece of the anterior parietal lobe belonged to the frontal lobe. I have also designated in the figures the parts O and Cu, which belong to the occiput. To what extent that part which we have termed temporal lobe and the gyri designated T. 1 and T. 2 belongs to the parietal lobe, remains to be proven. That the greater portion of the anterior parts of the lower two super-sylvian gyri belong to the parietal lobe is certain; and it may be that the fissures which, in the Fox and Bear, for example, is usually called the Sylvian fissure (see S in Fig. I and Fig. III, Recapitulation) is in fact only its posterior ascending branch (S' in Fig. I of Introduction).

The ingenious successor to Leuret and Gratiolet will, in a measure, have to recede from his enthusiasm respecting the parietal lobe, while other anatomists will have to become more interested in it than they have hitherto been.

Broca is extraordinarily clear in his representations of the origin of the *fissura Sylvii* at the base and concerning its relations in animals. In general views his work is so rich, and detail so ingenious, that over and above its abundant explanations, every reader will derive real aesthetic enjoyment from it.

IV.

It is an important question whether, from an atypic skull, we are able to determine an atypic brain.

Let us consider the various sections of the longitudinal curve.

In typic skulls and brains the bregma lies about 4.5 Cm (1 3-4 inch) anterior to the *sulcus centralis* (c) and the intersection of the sagittal and lamboidal sutures is at the upper terminus of the parieto-occipital fissure. The question is, does this relation remain the same in atypic skulls and brains?

To the present, science can give no response.

Let us suppose that the posterior fontanelle should always correspond with the superior end of the *fissura-paristo-occipitalis*.

In the specimens which we have presented, there is the abnormality—at least it occurs with unusual frequency—that the cerebellum is not covered by the occipital lobe.

Let us see if we can discern this condition from the skull.

It would be imagined *a priori*, that with the abnormal brain (in this respect) the condition must be manifest by the curve extending from the posterior fontanelle to the *prominentia occipitalis maxima* (upper occipital curve). It must be observed, however, that the curve of the *prominentia occipitalis maxima* is generally a circular one, and that therefore for the purpose of measuring this curve (upper occipital curve) it affords somewhat of an arbitrary point which is more especially true in case of a flat occiput. Besides this, the *prominentia occipitalis maxima* does not of a certainty correspond to the intersection of the *eminentia cruciata*, which last also exercises a considerable influence in deciding the true extent of the occipital curve of the brain.

Arranging the skulls according to the size of this curve, we have the following order:

1, Beczar,	. . . . . .	1.5
2, Budimcie,	. . . . . .	2.0
3, Faczuna,	. . . . . .	3.1
4, Pantalić,	. . . . . .	3.4
5, Sinka,	. . . . . .	3.4
6, Petriczewicz,	. . . . . .	3.7
7, Proketz,	. . . . . .	4.4
8, Loksik,	. . . . . .	4.5
9, Madarasz,	. . . . . .	5.1
10, Perndinacz,	. . . . . .	5.5
11, Mia,	. . . . . .	5.6

Let us exclude Beczar, in whom the cerebellum was entirely covered from the base of the occipital lobe down, and Budimcie, with whom the small measurement indicates microcephaly and which therefore cannot be ranked as a case of non-covering.

Of the remaining nine, there are three without covering (MADARÁSZ, PANTALIĆ, PROKETZ), three barely covered (PERNDINACZ, SINKA, LOKSIK) and three plentifully so, (FACZUNA, MIA, PETRICZEWICZ).

From the comparison in this collection, it is to be seen that NO conclusion can be drawn from the size of the "upper occipital curve" of the skull as regards the extent to which the cerebellum within is covered by the occipital lobe.

The next question is, does an abnormal proportion of the cranial-parietal curve indicate an abnormal development of certain parts of the brain?

The reply must at all events be ambiguous. Within the range of the cranial-parietal curve there is included a portion of the anterior lobes, then the central lobes, and lastly the parietal lobes. The question therefore is: in a contracted parietal curve of the skull, are the entire brain-parts just enumerated, dwarfed, or is it confined to single parts? and to what parts?

In this direction our Fifth observation is of interest. (SINKA.)—In his case the non-development of the bony parietal curve is strongly marked. The man was a bank-note counterfeiter, and this class of criminals is noted for geniality and never for intellectual poverty. As we have every reason to locate intelligence in the frontal cerebral curve and to regard the *gyrus centralis anterior* and its anterior neighboring parts as the *psychomotor centrum*, it might be expected, *a priori*, that a connection would be found between the non-development of the bony parietal curve and an aplasia of the parietal lobes.

A glance at the photographic plate V, shows in fact a highly dwarfed condition of the first parietal lobe, by an operculose formation, and this on both sides.

The aplasia of the parietal lobe in the robber and thief, PROKETZ, is less pronounced (Observation XV).

The incomplete development of the cranial-parietal curve in epileptics—and such is often the case—might be connected with aplasia of the *gyrus centralis anterior*.

The verification of asymitries of the base of the skull and of the skull entire, seems also of importance. Still even from this there could, as yet, be no conclusions drawn respecting partial aplasias.

For the present the skull and the brain must be studied as parallel lines much more than as lines exactly superimposed.

I have selected certain measurements of the brain in order to establish at least relative references.

The greatest longitudinal diameter ("L") of each hemisphere, in proportion to the longitudinal arch-curve ("Hemispheric arch") may be of the greatest importance. First comes the measurement of the "anterior" (frontal) curve, from the apex of the frontal lobe to that point where the *sulcus centralis*, if extended, would cut the hemispheric arch; the "middle" (parietal) curve extends from this point to the upper end (medial surface) of the *fissura parieto-occipitalis* and the "posterior" (occipital) curve extends from this to the *prominentia occipitalis*. These curves may afford good measurements to ascertain the relative development of various parts of the brain. In our observations they are of little value because they have been made on dried brains, and the shrinkage of the brains *in toto* as well as in different parts, varies greatly in the different subjects. The measurements, to have value, must be made upon either fresh brains or those slightly hardened. The first method might be difficult and the latter would require preparatory experience.

In conclusion, I will offer a few remarks upon the hitherto-existing Criminalistic Method.

The plan pursued has been the most injudicious possible. Crime is a psychological act of the criminal, and the criminal, therefore, is the first object for study. In regard to this,— the very chief object—the teacher, the accuser, the defender and the judge, all have to the present bestowed too little care.

The most competent judges of criminals are empiric policemen, and above all thorough prison officers. Brief conversations with Director Tauffer, and the honor of a longer interview with the Ex-General-Inspector of Austrian prisons, Baron von Hye, who, as a theorist and a practicalist has equal prominence and reputation, have taught me more concerning the state of the criminal world than would an entire library of books.

In order to acquire and spread abroad sensible views, and before all, that it may be clearly ascertained whether and how criminals can be corrected, and how society can best be protected from the scourge of crime, it will be necessary for scientific criminalists to adopt the methods of naturalists. It will be especially requisite to have institutions for purposes of

RECAPITULATION IV.

observation and teaching, and these must be established in capital cities where prisons and the higher executive departments of justice are located, and above all they should be confided to educated and scientific investigators.

For the future, judge, defender, and prosecutor, to be acquainted with the subject of their labors face to face will certainly be more to the purpose than to accept the multitude of false psychological theories which have been enunciated by teachers from far back, prehistoric times, and which are generally taught even in the present day.

If haply in one of these institutions some future professor —somewhat mocking Lombroso,—exclaims to his pupils: " The heads of criminals must not be too large or too small, not too broad or too narrow, not too high and not too low, what then should they be ?" The proper reply would at once echo back: "Yes indeed, these heads may be too large or too small, too narrow or too broad, too high or two low; they may be atypic." For atypy predisposes to mental disturbance, to epilepsy and to psycho-physical abnormalities of all sorts; or it may be significant of cerebral disease. But for the criminalist it is necessary that his head should not be hollow, in order that he may be able to follow investigations originating from other sources than himself; and in order that he may not be incited to slander other investigators because he himself is incapable of grasping principles, it is necessary that his head should not be evil.

www.ingramcontent.com/pod-product-compliance
Lightning Source LLC
Chambersburg PA
CBHW032152160426
43197CB00008B/880